Tropical
Deforestation

The Tropenbos Foundation was established in July 1988 to continue and expand the International Tropenbos Programme, set up by the government of the Netherlands in 1986.

The main objectives of the Foundation are:

- to contribute to the conservation of tropical rain forests and promote their wise use, by generating knowledge and developing methodologies;
- to involve local research institutions in the above objectives and to increase their capacity in this regard.

The Tropenbos Foundation formulates, organizes and finances objective-oriented, multidisciplinary research programmes. In close cooperation with research institutes and governments in a number of tropical countries and in the Netherlands, several major research sites have been established. At present, sites are operational in Colombia and Guyana (South America), Côte d'Ivoire and Cameroon (Africa), and on Kalimantan, Indonesia (South-East Asia).

Tropical Deforestation

A Socio–Economic Approach

CJ JEPMA

EARTHSCAN

Earthscan Publications Ltd, London

First published in the UK in 1995 by
Earthscan Publications Limited

A catalogue record for this book is available from the British Library

ISBN 1 85383 238 3

Printed and bound by Biddles Ltd, Guildford and King's Lynn
Cover design by Elaine Marriott

For a full list of publications please contact

**Earthscan Publications Limited
120 Pentonville Road
London N1 9JN
Tel: 0171 278 0433
Fax: 0171 278 1142**

Earthscan is an editorially independent subsidiary of Kogan Page
Limited and publishes in association with WWF-UK and the
International Institute for Environment and Development.

CONTENTS

TABLES, FIGURES AND BOXES

Tables

Figures

Boxes

ABBREVIATIONS AND ACRONYMS

AFREST	Rest of Africa
AFTROP	Tropical Africa
AIC	Advanced Industrialized Country
AREA	Australian Resources and Environmental Assessment
ASREST	Rest of Asia
ASTROP	Tropical Asia
BFL	Basic Forestry Law
bn	billion
CBA	Cost Benefit Analysis
DEVPA	Developed Pacific
EC	European Community
ECE	Economic Commission for Europe
EIA	Environmental Impact Assessment
EURCPL	Europe (formerly) Centrally Planned
EUROP	Western Europe
FAO	Food and Agriculture Organization of the United Nations
FB-SSI	Farm Based Small Scale Industries
GDP	Gross Domestic Product
GNP	Gross National Product
ha	hectare
HDI	Human Development Index
ICRAF	International Council for Research in Agroforestry
IDIOM	IDE Deforestation and the International Order Model
IIASA	International Institute for Applied Systems Analysis
INDUS	Industry
ITTO	International Timber Trade Organization
IUCN	International Union for the Conservation of Nature and Natural Resources
LAREST	Rest of Latin America
LATROP	Tropical Latin America
LEEC	London Environmental Economics Centre
LEISA	Low External Input Sustainable Agriculture
LP	Linear Programming
LPG	Liquified Petrol Gaz
MJ	Megajoule
m	million
NATPRD	Natural Products
NEI	Nederlands Economisch Instituut
NIC	Newly Industrializing Country
NORAM	North America
OECD	Organization for Economic Co-operation and Development
PPP	Purchasing Power Parity
R&D	Research & Development

SARUM	System Analysis Research Unit Model
SBH	Stichting Bos en Hout
SERVIC	Services
TFAP	Tropical Forestry Action Plan
TFR	Total Fertility Rate
TPTI	Indonesian Selective Cutting and Replanting System
UN	United Nations
UNCD	United Nations Conference on Development
UNCED	United Nations Conference on Environment and Development
UNDP	United Nations Development Programme
UNESCO	United Nations Educational, Scientific and Cultural Organization
URGD	Urban/Rural Growth Difference
USA	United States of America
USSR	Union of Socialistic Soviet Republics
WB	World Bank
WRI	World Resources Institute
WWI	World Watch Institute
WWF	World Wildlife Fund

PREFACE AND ACKNOWLEDGEMENTS

This report deals with the issue of global tropical deforestation from a socio-economic point of view. The tropical forests are disappearing at an increasing rate as a consequence of both direct causes such as agricultural encroachment and logging, and indirect causes associated with general socio-economic conditions. Policies (or policy changes) in each of these areas will affect deforestation rates and have to be included in the analysis.

The research carried out by the International Development Economics (IDE) foundation and the Department of Economics of the University of Groningen (The Netherlands) in the framework of the Tropenbos programme consists of an analysis of economic factors and developments associated with deforestation in the humid tropics, resulting in an evaluation of the effect of various policies on deforestation rates. Therefore a global policy simulation model, named IDIOM, has been developed. This model is suited to evaluate the impact of various sets of policy options on deforestation rates in the humid tropics. The final model integrates three main modules: a general development module describing the main interlinkages in the world economy (SARUM); a module focusing on the various aspects of tropical timber production and trade, and their relationship with deforestation (TROPFORM); and a module dealing with the various aspects of agricultural land use in the tropics. This report discusses the model structure, and presents the model results of a base scenario and various policy scenario simulations.

The study was carried out during the period September 1989 to February 1993 by a research team under my supervision. Research fellows were Marcel Blom, Conny Hoogeveen and Erik Jansen. They are all thanked for their contributions and effort, especially Conny for her continuous dedication to the project. In addition many thanks are due to Dr. Kim Parker (University of Kent) with whom a true partnership developed and who made a significant contribution to the research activities during the deforestation project, especially related to Part III of this book. Finally, I would like to express a word of thanks to research assistants R. de Bruyn, M. Dulleman, A.R. Gigengack, G. Manganelli and J.M. Schippers. A list of background documents to this report is presented in Appendix 6.

Catrinus J. Jepma
Groningen
June 1995

PART I

TROPICAL RAINFORESTS AND DEFORESTATION

INTRODUCTION

The concern about the present speed of global tropical deforestation and tropical forest losses that have already occurred, especially during the last decades, is so widespread that the issue is now commonly perceived as a major international one. Part of the concern explicitly highlights the potential damage for people, both current and future generations; another area of concern merely focuses on the vulnerability of the ecosystems. Often, however, these two major concerns about tropical deforestation are seen as going hand in hand.

One has to admit that tropical deforestation may have economic benefits for various groups involved, at least in the short term, but it can also have serious economic repercussions. One of the main economic benefits from tropical deforestation is the increased land area that becomes available for economic use, such as for shifting cultivation, plantation or agroforestry or cattle ranching. If the deforestation is due to logging, another benefit is the availability of timber.

The costs, however, are first of all related to lost opportunities, through the production of non-timber products (e.g. through the traditional hunting-gathering tribes, but also through a more 'commercial' approach), or through recreation (ecotourism). After all tropical forests have built up the world's largest biomass per hectare and contain a great diversity of plant and animal species. In addition, deforestation can have an adverse impact on the fertility of the area through soil erosion and run-offs. The costs in downstream areas can be sizeable indeed, because forests have a twofold buffer activity; first, the tree canopy intercepts the rain, and second, the humus and the roots absorb and recycle water. Loss of these functions results in rivers coming from deforested lands flooding excessively after a downpour, but quickly running dry thereafter (e.g. the Himalaya slopes versus the rivers in the Ganges plain and Bangladesh).

Another negative factor in deforestation is that the local climate can be adversely affected, often in an unquantifiable way. Since evaporation greatly increases the humidity in the vicinity of a forest, the taller the forest the more water it puts into circulation. This removes from a local climate the extremes of heat, cold and drought. If the supply of water vapour dwindles, the zones which suit certain crops soon shift and/or become narrower (especially in the large continental masses of Africa and Latin America).

A third cost factor is due to the loss of the world's genetic materials, half of which are probably located in the tropical forests. Precisely because the economic losses that may be involved with the extinction of various plants and animals remain for the most part unknown, these losses are generally perceived as a matter of great concern. Moreover, tropical rainforests, once destroyed, are almost impossible to restore, for the tropical rainforest perpetuates itself in 'cyclical regeneration'.

Ecologically, therefore, it is a stable environment, with the maximum number of species possible under the circumstances: any external influence can only result in impoverishment. Ironically there is nothing to manage in such a system; the only way to preserve it is to protect it from interference of any kind.

Finally, the relationship between global tropical deforestation and the global climate issue, the greenhouse effect in particular, has become a major concern. Because tropical deforestation to a large extent takes place through slash-and-burn techniques, and because the CO_2 content of the replacing biomass is less than that of the original vegetation, there is a net CO_2 emission involved with the disappearing tropical forest. Although there is considerable variation in estimates on how this effects the overall CO_2 emissions, roughly between 7–30 per cent (Mors, 1991, p. 81), it is in any case clear that this cost component of deforestation is serious indeed. It may involve not only the negative impact of climate alterations, such as shifts in the rainfall patterns, in moisture, or the mean air and/or sea temperature, but especially the exponential expansion of extreme weather conditions due to small shifts in mean values of climate variables.

1

TROPICAL RAINFORESTS

GEOGRAPHICAL DISTRIBUTION OF TROPICAL RAINFORESTS

The total area of the globe where the conditions are fulfilled for tropical rainforests to survive can be estimated as some 1046 million hectares (Hagget, 1979); according to WRI (1990, Table 19.1, pp.292-3) the tropical forests currently account for 42 per cent of the world's total forest area. The major share of this tropical forest area is situated around the equator, either in Latin America and the Pacific, or in Asia and Africa. The main countries where the conditions for closed forests of the permanently humid tropical regions are fulfilled are Brazil, Zaire and Indonesia, with a forest area of 34, 12 and 8 per cent respectively of the world total; the remainder is situated in some 40 other countries.

According to Jacobs (1988, pp.2-6) the following physical conditions have to be fulfilled to enable the natural development of a tropical rainforest:

- A relatively continuous tropical temperature, with a small amplitude both through the day and throughout the year. This condition generally holds for latitudes between 23°30´N and 23°0´S, at least if the land is below 1000m altitude.
- At least 1800mm rainfall per annum, evenly distributed throughout the year.

These conditions have proven to be necessary but not always sufficient for the natural development of tropical forests.

It is obviously a matter of definition to assess the precise boundaries of the rainforests, especially if these forests gradually transit into non-evergreen forests. The practices that have led to deforestation have exacerbated the difficulties of defining the tropical forest area both in qualitative and quantitative terms. However, the main range of a tropical rainforest is well-known. Jacobs (1988, pp.3-4) distinguishes the following main regions:

1) **The American rainforest region.** On the South American continent three rainforest subregions can be distinguished:
    - Amazonia, by far the largest and still mainly intact;
    - a second region west of the Andes and north of the equator extending intermittently to Mexico. This region has already suffered badly;

- a third and smallest region consisting of a narrow strip along the Atlantic coast of Brazil between 14º and 21ºS; only a few remnants of it still exist.

2) **The Malaysian rainforest region or Southeast Asian region.** This region coincides almost entirely with the former Malaysian Archipelago (the Papua New Guinea and Irian Jaya island included), to which the Malay Peninsula belongs botanically; the true rainforest is almost absent from continental Asia, except in Malaya, southern-most Thailand and Southwest Cambodia.

   In this region two closely related nuclei can be distinguished. First, Borneo-Malaya-Sumatra and The Philippines, with outliers in the Andaman Islands, Sri Lanka and the poorer forests of Southwest India. Second, the Papua New Guinea and Irian Jaya island, including the less species-rich forests on the islands to its west and east, with an outlier along the eastern coast of tropical Australia, where some pockets of rainforest have survived. Forests in Malaya, Sumatra, Borneo and The Philippines have been heavily depleted in recent times: the Papua New Guinea and to a lesser extent the Irian Jaya forests thus far have remained largely intact.

3) **The African rainforest region.** This region consists of a number of subregions, all partly destroyed, both along the Atlantic coast between ca. 10ºN and 5ºS, and in the Congo Basin stretching east to the mountains. In addition, some outlier pockets in East Africa may be regarded as more or less true rainforests.

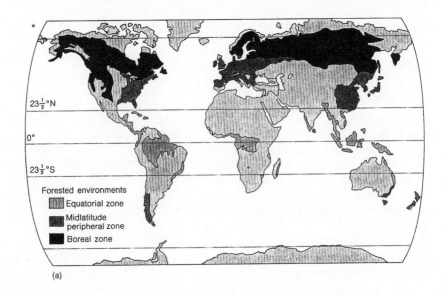

(a)

Source: Hagget (1979)

Figure 1.1 Main environmental regions

ESTIMATES OF FOREST RESOURCES AND DEFORESTATION

The size of the tropical forest resources and the amount of deforestation obviously depend on how they are defined. Since there is no consensus about the correct definition of both concepts among the various experts and institutions involved, a definite assessment of the size of deforestation processes cannot be made, even if the registration techniques are perfect, which they are not.

The extreme points of view with respect to forest resources and deforestation can be traced down to different angles from which the tropical forests are viewed. On the one hand one distinguishes what could be labelled as the environmentalists' point of view, according to which the tropical forest should be considered primarily as an unique ecosystem that should be preserved in all its biological diversity. On the other hand one can distinguish the forestry sector point of view. According to the latter, the tropical forest is recognized as being a vulnerable ecosystem, but should be valued equally on the basis of the economic value of its resources, e.g. through the exploitation of its timber or otherwise.

These different points of view, i.e. those of the ecologist vs the economist, pervade the whole debate on the modalities, size and scope of deforestation, and also on how best to deal with the issue through policy measures. The above caveat should also be borne in mind when reviewing the various definitions of deforestation, and even impact upon the forest resources estimates.

Box 1 Main peculiarities of the rainforest: a botanist's point of view
(Jacobs, 1988, pp.4–6)

1) They contain the largest number of species, of animals and plants, of any known ecosystem;
2) they occur generally on poor soils;
3) in comparison to other forests, their potential for utilization is greatest in terms of quality, but smallest in terms of quantity;
4) they cannot easily be cropped in quotas;
5) they contain huge capital assets in the form of timber which, unlike non-timber products, generally cannot easily be harvested without serious ecological damage;
6) they cannot be 'managed' without the loss of a large number of species;
7) they are unusually fragile, and, once damaged, do not recover, or recover too slowly for any human planning;
8) nearly all of them are situated in developing countries with shaky economies, and little or no power to implement laws on land use, if and where such laws exist;
9) they are now being logged rapidly by mechanical means since for a decade machinery has been available;
10) yet they are so complex that even their main ecological features are poorly known even to professional decision-makers and misunderstandings about them abound.

A predominant view in the literature which can be associated with the above mentioned forestry sector point of view – and which is reflected in the deforestation data of both FAO and WRI – considers deforestation as a complete clearance of tree formations (closed or open) and their replacement by non-forest land uses. Environmentalists have criticized this definition. Barraclough and Ghimire (1990, p.7) for instance, argue firstly that the definition 'does not view as deforestation the removal of plant associations not classified as forest'. Secondly, and most fundamentally, they criticize the definition because serious forest damage caused by excessive logging, wood gathering for both domestic and commercial purposes, fire and livestock grazing is not considered as deforestation unless it results in total conversion of forests to other land uses. Biologists, ecologists and conservation agencies therefore tend instead to consider deforestation as the degradation of 'entire forest ecosystems' involving wildlife species, gene-pools, climate and biomass stocks.

An example of the alternative view on how to define deforestation is Myers' definition. According to him (1991, p.4) deforestation not only involves: 'the complete destruction of forest cover through clearing for agriculture of whatever sort' (cattle ranching, smallholder agriculture, whether planned or spontaneous, and large-scale commodity crop production through, for example, rubber and palm-oil plantations). In addition, he feels that the definition should be extended to cases of serious degrading of the tropical forest ecosystem:

> there are certain instances too where the forest biomass is so severely depleted – notably through the very heavy and unduly negligent logging of dipterocarp forests in Southeast Asia, resulting in the removal of or unsurvivable injury to the great majority of trees – that the remnant ecosystem is a travesty of natural forest as properly understood. Decline of biomass and depletion of ecosystem services are so severe that the residual forest can no longer qualify as forest in any practical sense of the word. So this particular kind of over-logging is included under the term 'deforestation'.

Some of the leading organizations advancing this view are the WWF and the IUCN. (With respect to the definition of deforestation, see also Barraclough and Ghimire, 1990, p.7.)

It should be clear beforehand that the difference in the definition as just described not only affects the measurement of the forest resources but also of annual deforestation. It seems fair to presume that Myers' definition involves less area to be labelled as tropical forest area than does the FAO/WRI approach. At the same time data on deforestation can be expected to be upward biased through the use of the more extensive definition suggested by Myers. So if both aspects are combined, deforestation data according to Myers may not only become larger through the definition of deforestation itself, but also due to the fact that deforestation processes are related to a smaller area defined as tropical forest. This expectation seems to be supported by some broad global estimates of average annual deforestation as presented in the past:

- FAO 0.8 per cent during 1981-1990 (FAO, 1993);
- WWF 1.15 per cent (Cross, 1988);
- Myers 1.8 per cent by 1989 (Myers, 1989).

A similar point of view is reflected in more recent data on tropical forests as presented in Tables 1.1a-1.c. These tables reflect our grouping of country data both on forest area and on deforestation as provided by WRI, FAO, and Myers.[1] The data from both sources are not directly comparable due to the difference in tropical forest and deforestation definitions (see above) but also as a result of a difference in timing of the data, and in the country coverage. Nevertheless, the different main outcomes with respect to deforestation are striking: Myers' data point at a speed of deforestation which is almost double the WRI estimate.

Table 1.1 Forest resources and deforestation in the humid tropics in the 1980s
a) Tropical Latin America (in 1000 ha unless indicated otherwise)

	Forest area estimates			*Average annual deforestation*					
	WRI[1]	FAO[2]	Myers[3]	WRI		FAO		Myers	
				1000 ha	%	1000 ha	%	1000 ha	%
Total/ average	785,875	823,500	389,950	10,918	1.4	5274	0.6	6650	1.7
Bolivia	66,760	49,317	7000	117	0.2	97	1.2	150	2.1
Brazil	514,480	561,107	220,000	9050	1.8	3671	0.6	5000	2.3
Colombia	51,700	54,064	27,850	890	1.7	367	0.7	650	2.3
Ecuador	14,730	11,962	7600	340	2.3	238	1.8	300	4.0
Guyana[4]	18,695	18,416	41,000	3	0.0	18	0.1	50	0.1
Peru	70,640	67,906	51,500	270	0.4	271	0.4	350	0.7
Surinam	15,000	14,768	na	3	0.0	13	0.1	na	na
Venezuela	33,870	45,960	35,000	245	0.7	599	1.2	150	0.4

Sources: WRI (1990), Table 19.1, pp.292-293; FAO (1993), Tables 4a-4c, annex 1; Myers (1991), Table 1, p.6.

Notes:
1) Total area early 1980s (closed and open forest).
2) Forest area 1990.
3) Present extent of forest cover.
4) Estimate Myers is the sum for French Guyana, Guyana and Surinam.

Because the FAO data are the most recent and represent the total global forest area fairly completely, it may be useful to use them as a basis for a representation of the broad regional distribution of the forest area and deforestation. Figures 1.2 and 1.3 reflect the regional distribution of tropical forest area, and the contribution of the main countries to global

---

[1] The country selection is based on their role in tropical timber production. The following analysis will focus on the same group of countries as much as possible.

Table 1.1 b) Tropical Africa

	Forest area estimates			Average annual deforestation					
	WRI	FAO	Myers	WRI		FAO		Myers	
				1000 ha	%	1000 ha	%	1000 ha	%
Total/ average[1]	463,928	349,239	52,200	2430	0.5	2616	0.7	1580	1.0
Angola	53,600	23,074	na	94	0.2	174	0.7	na	na
Benin	3867	4947	na	67	1.7	70	1.3	na	na
Cameroon	23,300	20,350	16,400	190	0.8	122	0.6	200	1.2
Central Africa	35,890	30,562	na	55	0.2	129	0.4	na	na
Congo	21,340	19,865	9000	22	0.1	32	0.2	70	0.8
Côte d'Ivoire	9834	10,904	1600	510	5.2	119	1.0	250	15.6
Eq. Guinea	1295	1826	na	3	0.2	7	0.4	na	na
Gabon	20,575	18,235	20,000	15	0.1	116	0.6	60	0.3
Ghana	8693	9555	na	72	0.8	137	1.3	na	na
Guinea	10,650	6692	na	86	0.8	87	1.2	na	na
Guinea- Buissau	2105	2021	na	57	2.7	16	0.8	na	na
Kenya	2360	1187	na	39	1.7	7	0.6	na	na
Liberia	2040	4633	na	46	2.3	25	0.5	na	na
Madagascar	13,200	15,782	2400	156	1.2	135	0.8	200	8.3
Nigeria	14,750	15,634	2800	400	2.7	119	1.2	400	14.3
Senegal	11,045	7544	na	50	0.5	52	0.7	na	na
Sierra Leone	2055	1899	na	6	0.3	12	0.6	na	na
Tanzania	42,040	33,555	na	130	0.3	438	1.2	na	na
Togo	1684	1353	na	12	0.7	22	1.5	na	na
Uganda	6015	6346	na	50	0.8	65	1.0	na	na
Zaire	177,590	113,275	100,000	370	0.2	732	0.6	400	0.4

Sources: WRI, 1990, Table 19.1, pp.292-293; FAO, 1993, Tables 4a-4c, annex 1; Myers 1991, Table 1, p.6.

Notes:

1) Total for the area only relates to the number of countries for which data were specified.

deforestation, respectively. Firstly they show the overwhelming role played by Brazil, a country estimated by FAO (1993) to be responsible for about 40 per cent of the tropical forest area, but also for about one third of the global deforestation. Secondly, Figure 1.2 suggests that over half of the tropical forest resources are located in three countries only: Brazil, Zaire and Indonesia. This raises the question of whether a global rather than a national preservation strategy should be the first

Table 1.1 c) Tropical Asia

	Forest area estimates			Average annual deforestation					
	WRI	FAO	Myers	WRI		FAO		Myers	
				1000 ha	%	1000 ha	%	1000 ha	%
Total/ average[1]	272,826	234,039	194,100	2751	1.0	3219	1.4	4200	2.2
Indonesia	116,895	109,549	86,000	920	0.8	1212	1.0	1200	1.4
Malaysia	20,996	17,583	15,700	255	1.2	396	2.0	480	3.1
Philippines	9510	7831	5000	143	1.5	316	3.3	270	5.4
Thailand	15,675	12,735	7400	397	2.5	515	3.3	600	8.4
Kampuchea	12,648	na	6700	30	0.2	na	na	50	0.8
Lao Dem. Rep.	13,625	13,173	6800	130	1.0	129	0.9	100	1.5
Vietnam	10,110	8312	6000	173	1.7	137	1.5	350	5.8
Myanmar	31,941	28,856	24,500	677	2.1	401	1.3	800	3.3
Fiji	811	na	na	2	0.2	na	na	na	na
Papua New. Guinea	38,175	36,000	36,000	23	0.1	113	0.3	350	1.0
Solomon Isl.	2440	na	na	1	0.0	na	na	na	na

Sources: WRI (1990), Table 19.1, pp.292-293; FAO (1993), Tables 4a-c, annex 1; Myers (1991), Table 1, p.6.

Notes:

1) Total for the area only relates to the number of countries for which data were specified

alternative one should aim for: if the handful of local governments of the main countries containing the tropical forests on their territory would develop their policies towards a more conservationist attitude, much could be achieved already by global standards. Thirdly, the impression that a limited set of countries is involved with the bulk of the tropical forest deforestation problem is reconfirmed by Figure 1.3. This figure illustrates that, according to the above FAO data, over 50 per cent of the annual deforestation takes place in four countries only: Brazil (33%), Indonesia (10.9%), Zaire (6.6%) and Venezuela (5.4%).

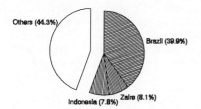

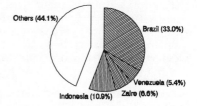

Figure 1.2 Share in forest resources (FAO)   **Figure 1.3** Deforestation (FAO)

Methods to estimate deforestation

Except for the conceptual differences vis-à-vis the assessment of deforestation processes, differences in the various data can also be attributed to the estimation procedure and the assessment technique. The FAO data, for instance, are based upon a set of national estimates of forest resources. For its 1990 assessment, FAO combined information obtained from various primary sources ranging from case studies and national statistics to remote sensing data. Yet the FAO approach has not remained without criticism. WRI (1990, pp.105–106) for instance argued that: 'some of the ground-based surveys relied on interpretations of old maps . . . literature surveys, and extrapolations from outdated inventories.' Others also pointed at the more fundamental limitation mentioned earlier that FAO data usually mainly focused on timber resources only.

In addition some more general remarks on forestry data through remote sensing are in order. It is generally recognized that despite ongoing improvements in remote sensing techniques both with respect to data collecting and interpreting, remotely sensed data are still not a panacea. Even state-of-the-art satellite images have their limitations. Some can distinguish only broad classifications of vegetation, while others provide greater resolution but a smaller scale, so that the cost of obtaining and interpreting coverage can be prohibitive; in some cases smallscale satellite imagery is even forbidden for security and military reasons. In addition, any space or airborne system can be foiled by the frequent cloudiness of the tropics, and ironically, by the smoke from the fires used to clear the forests. Finally, and most importantly, because deforestation is such a highly political issue, even the most impeccable data are subject to considerable differences of interpretation depending on the various responsibilities of the institutions involved.

One of the odd consequences of the continuous improvements in the quality of tropical deforestation data collection, as well as in data interpretation (e.g. through more precise definitions), is that in comparing the recent data with earlier ones, one becomes less certain about the true causes of the registered tropical forest decline. While undoubtedly most of the new studies benefited from the progress made over the past decade in collecting and interpreting remote-sensing data, it is, for instance, possible that the newer FAO studies' outcomes are not comparable to FAO's original assessment, precisely because of:

> the improved survey techniques, differences of the forest types included, and differences in defining deforestation and conversion. It has been suggested that what seems as a soaring deforestation rate is simply caused by the fact that the new surveys are more accurate and thus reveal old deforestation that had not previously been detected. (WRI, 1990, pp.105–106).

Myers' deforestation estimates have been equally criticized, however, for being soft and for their rather eclectic and therefore speculative methodological base (see for instance Sedjo and Clawson, 1984, and

Barraclough and Ghimire, 1990). With respect to the latter, Myers himself (1991, p.5) indicated:

> while it has drawn on the author's 20 years of travel and work throughout the biome, it has relied primarily on an extensive survey of the recent professional literature . . .. Further, the survey has entailed copious correspondence . . . plus numerous follow-up enquiries. The survey's findings have depended heavily, moreover, upon remote sensing data . . . backed up by ground truth checks.

Discussion about deforestation measurement techniques, although highly valuable in itself, should not overshadow the main thrust of the problem, namely that irrespective of the deforestation assessment approach, all sources clearly point out the alarming rate of annual deforestation. The various sources also make clear that deforestation rates do differ substantially across the various tropical regions, and that in many countries, such as Brazil, Colombia, Ecuador, Côte d'Ivoire, Nigeria, Thailand, Malaysia and The Philippines, to mention but a few, deforestation is or has been spectacular indeed.

DECOMPOSITION OF DEFORESTATION PROCESSES

One of the first questions that arises with respect to the various deforestation processes is what their main contributing factors are, and more particularly to what extent each of the various factors can be held responsible for the resulting tropical forest decline. Various studies have been carried out to try to broadly classify the deforestation process into the various distinct underlying factors. Despite the undisputable merit of those approaches for the design of the proper set of policy measures, two major caveats should be kept in mind.

Firstly, the various underlying factors are not totally without any mutual correlation. To give an example, there is some evidence that commercial logging activities can induce shifting cultivators to further penetrate into the forest due to its increased accessibility; the same can equally apply to the impact of infrastructure for mining or oil exploration upon labour mobility. In the Ecuadorian Amazon region, for example, the pattern of agricultural colonization of the region almost entirely coincides with the oil exploration area (see Hicks et al., 1990, maps IBRD 22048 and 22052). Such interlinkages obviously make it hard to draw precise dividing lines between the deforestation that can be attributed to commercial logging, mining or oil exploration activities on the one hand, and the mobility of shifted cultivators on the other hand.

Secondly, a breakdown of the various factors that can directly be held responsible for the actual deforestation, still does not provide the ultimate answer with respect to the question of what the contribution of the various (equally interrelated) **underlying** causes, such as population pressure, local government policy or labour migration, is. In any case this requires substantial additional analysis (see also pp.18–23).

Nevertheless, breaking deforestation down into its underlying factors can provide us with a highly valuable starting point. For instance, it enables us not only to demystify some popular misconceptions about deforestation, but also to provide a realistic framework for a tropical forest research and policy agenda.

The next three factors[2] may serve as an illustration. Against the background of the former remarks it is clear that the outcomes should be interpreted as orders of magnitude rather than as precise assessments.

According to the WRI (1990) there are three main causes of deforestation:

1) Permanent conversion of forest into agricultural land. These processes further expanded due to population pressure, whereby traditional slash-and-burn techniques, through shorter fallow periods, lead to land degradation.

2) Logging. In practice, the amount of sustainable forestry in the tropics is negligible; the usual pattern is that of damage of non-target trees and the opening of virgin forest areas for agricultural settlement.

3) Demand for fuelwood, fodder and other forest products.

---

[2] Another classification of deforestation processes, based on a literature survey and data from FAO and other sources is presented by Otto (1990). In this the world averages of the direct causes of deforestation have been assessed as follows:

a) **Burning down of tracts of forest** (responsible for 63 per cent of deforestation). One reason for this kind of degradation of the tropical forest is the well-known population pressure induced search for land for short-term agricultural purposes by poor landless farmers (squatters). A second reason is the official settlement schemes that take place, for instance in Brazil and Indonesia. A third reason is that, in Latin American countries, these poor squatters are sent by rich landowners to burn down the rainforest in exchange for the possibility of farming this infertile land for a few years. The logging or burning down should be interpreted, according to these governments, as improving the land. On that base the improver becomes entitled to or even owner of the land (Repetto, 1990, p.21 and Poore 1989, p.88).

b) **Reclamation of forest for the benefit of commercial agricultural plantations or mines** (responsible for 16 per cent of deforestation). This reclamation is often organized and subsidized by the government and is stimulated by a large international demand for minerals and agricultural products.

c) **Felling of the trees for firewood** (8 per cent). A fact of life that is often misunderstood is that by far the major part of the wood harvested in developing countries is not used for commercial purposes such as for timber exports, but is instead used locally, mainly for fuel. This share has been estimated at 70 per cent (Repetto, 1990, p.20) or even more. Indeed, over 2 billion people use firewood to cook their meals!

d) **Reclamation of the forest for the benefit of large scale stock farming** (6 per cent). The driving forces behind the cattle ranches are the large amount of subsidies these farmers receive; the great demand for cheap meat ('the hamburger connection'); and the use of the land as a speculative venture in countries with high inflation rates.

e) **Commercial exploitation of timber (logging) on behalf of domestic and foreign industrial use** (6 per cent). Export of logs and sawnwood is a source of foreign exchange receipts, and so it is stimulated by those domestic governments.

f) **Construction of infrastructure facilities** (1 per cent), such as roads and power plants.

As far as we could see, no emphasis has been attached by WRI to the contribution of these three causes.

Although there are regional exceptions, experts generally seem to agree that the first cause is by far the most important. This view seems also to be generally accepted in policy circles: witness, for instance, the text of an official white paper by the Netherlands' government of 1990 expressing its view on tropical rainforests issues. According to this text, extension of the agricultural area accounts for about 80 per cent of total tropical forest degradation (the Netherlands' government view on tropical forests, white paper 1990, annex 5, p.8).

Myers (1991) is somewhat more explicit about the contribution of the various factors to the deforestation process than WRI. According to his data the present annual deforestation amounts to over 140,000 km².[3] The main part of it, a little over 87,000 km², is, according to Myers, 'ostensibly due to slash-and-burn farmers, mainly shifted cultivators . . . (they) account for 61 per cent of all forest destruction, a proportion that appears to increase rapidly.'

Twenty–one per cent of the remaining 39 per cent can, according to Myers, be attributed to those commercial logging activities (in his view mainly in Southeast Asia) that amount to forest destruction. Although he believes that the commercial logger is currently affecting some 45,000 km² of new forest each year – much the same as ten years ago – roughly two thirds of this, or 30,000 km², is so heavy and negligent that it amounts to forest destruction.

Cattle ranching, almost entirely confined to Central America and Amazonia, according to Myers, causes 15,000 km² (11 per cent) of forest to be cleared each year – rather less than 10 years ago – and forest conversion to cash-crop plantations (oil palm, rubber, etc.) plus forest destruction for roads, mining and other activities of similar, relatively small-scale sort, amount to perhaps 10,000 km²/year (7 per cent). The total contribution to annual global deforestation of the three 'non slash-and-burn' categories therefore amounts to some 55,000 km².

In addition to the above 'classical' contributory factors, Myers points to some potential additional sources of deforestation. These will be very hard to quantify, however:

> There are still further possible sources of deforestation, these being atmospheric and climatic in nature. True, they are not likely to make their impact felt until early next century, whereas the factors described are already powerful causes of deforestation. But these additional factors could eventually prove to be some of the most harmful of all. They include region-scale climatic feedbacks, global-warming feedbacks, and acid rain, all of which could serve to induce pronounced depletion of tropical forests.
>
> (Myers, 1991, p.25)

Natural causes of deforestation like woodfires, droughts and cyclones were, although not insignificant (Jacobs, 1988, p.23), not included in the subdivision.

---

[3] This figure is slightly larger than his total deforestation estimate as presented in Tables 1.1a–c (122.5 km²), due to the country limitation in these tables.

The main *underlying* contributing factor, according to Myers, remains population pressure: the single largest agent of tropical deforestation is the 'shifted' cultivator (Myers, 1991, p.21). 'They are squeezed out of traditional farmlands, and head for the last unoccupied lands they are aware of, the forests.' Driven significantly by population growth and sheer pressure on existing farmlands (albeit often cultivated with only low or medium levels of agrotechnology, hence cultivated in extensive rather than intensive fashion), slash-and-burn farming is, according to Myers, the principal factor in deforestation in Colombia, Ecuador, Peru, Bolivia, Nigeria, Madagascar, India, Thailand, Indonesia and the Philippines, and probably also in Mexico, Brazil, Myanmar (Burma) and Vietnam. Populations of these cultivators are often increasing at annual rates far above the rates of nationwide increase. In Rondonia in Brazil the numbers of small-scale settlers have been growing at a rate that has surpassed 15 per cent per year for much of the period since 1975, whereas the population growth rate for the whole of Brazil has averaged only 2.1 per cent. Although exact figures are not available, it seems safe to assume that a large share of this population growth can be attributed to immigration. There are similar mass migrations into tropical forests, albeit at lower rates, in Colombia, Ecuador, Peru, Bolivia, Ivory Coast, Nigeria, India, Thailand, Vietnam, Indonesia and the Philippines. In all these instances, population growth is 'a significant if not the predominant factor in deforestation' (Myers, 1991, p.21).

Population pressure as an explanation for tropical deforestation seldom stands on its own. It shows itself in various patterns and is related to various socio-economic processes, such as poverty among peasant communities concerned, maldistribution of existing farmlands, inequitable land-tenure systems, inefficient agrotechnologies, low support for the subsistence-farming sector, lack of rural infrastructure, and faulty development policies overall.

Not in all countries, however, is population pressure an equally important, or even the main factor responsible for deforestation. Various tropical forest regions are sparsely populated, such as the island of New Guinea, or the countries of the Zaire Basin in central Africa, or the countries of the Guyanas, viz. Guyana, Surinam and French Guyana, and the western sector of Brazilian Amazonia. These cases differ fundamentally from those where population pressure is strong. Deforestation is much less an issue here, and is mainly due to logging activities. In other tropical forest regions, such as Central America, the impact of population growth is overshadowed by the spread of cattle ranching on the part of a relatively few large-scale cattle ranchers.

Amelung (1991, in LEEC 1992, Tables 3.2a and 3.2b), finally, in a study on the contribution of various sectors to deforestation shows that the extent to which a sector can be held responsible for deforestation is highly sensitive to how deforestation is defined. Amelung distinguishes between deforestation as defined by the FAO (a change of land use or a depletion of crown cover to less than 10 per cent), and other deforestation definitions, i.e. forest degradation and forest

Table 1.2 Sectoral shares in deforestation, forest degradation and forest modification (all major tropical countries, in %)

	Deforestation[1]	Degradation[2]	Modification[2]
Forestry	2	10	71
Agriculture	83	76	26
shifting cultivators	47		
permanent agriculture	36		
pastures	17		
permanent crops	3		
arable land	16		
Others	15	13	4
Mining (incl related industries)	na		
Hydroelectricity production	2		
Residual	13		

Source: Amelung (1991), in LEEC (1992), Tables 3.2a and 3.2b.

Notes:

1) Averages for the period 1981–1988;
2) Figures refer to the 1981–1985 period.

modification in which other changes of forest systems are also included. In Table 1.2 the sectoral contribution to deforestation depending on the definition employed is summarized. From the table it becomes clear that the less rigidly deforestation is defined, the more significant the role of the forestry sector becomes. Whereas the contribution of the forestry sector to deforestation in the strict sense accounts for a mere 2 per cent, this share rises to 71 per cent if a more sensitive definition of deforestation is used. For the agricultural sector the reverse holds: whereas the agricultural sector accounts for 83 per cent of all deforestation according to the FAO definition, this share falls to 26 per cent in a more comprehensive view on deforestation.

In conclusion, the general consensus seems to be that land use through shifting cultivation is the main factor responsible for tropical deforestation, mainly through burning down of tracts of forest. Only in the case where deforestation is very broadly defined as Amelung shows (see Table 1.2, 'Modification'), forestry activities are considered to be the main source of deforestation. In all other cases estimates of the contribution of the agricultural sector roughly vary between 60 and 80 per cent. Moreover, as will be discussed later on in the text, there are strong indications that the impact of this factor will further increase in the future.

TROPICAL DEFORESTATION AND LOCAL GOVERNMENT POLICIES

In the previous section, the policies of the local governments have not explicitly been mentioned as one of the contributory factors to tropical deforestation, because by definition the decomposition merely tackled the directly contributing rather than the underlying factors, such as the various policy measures that affect deforestation. Some authors claim in their analysis, however, that local policies can make a world of difference when it comes to deforestation processes, or can even be viewed as the main underlying factors responsible for the phenomenon.

Repetto (1990) for instance argued that the inefficient commercial logging operations and the conversion of forested areas to cattle ranching and agriculture leading to current rapid deforestation rates are largely the result of a failing government policy: 'deforestation is largely the consequence of poor stewardship, inappropriate policies and inattention to significant social and economic problems whose true locus is outside the forest sector.' Repetto (1989) distinguishes between resource degradation as a consequence of market failures which governments are unable or unwilling to correct, and policy induced market distortions:

> Some incentive problems arise from market failure. For example, people borrow against the future by destroying renewable resources because they lack options. They persist in using inappropriate technologies because they lack the knowledge and resources to adapt. They ignore future consequences because institutions deny them a secure stake in the future yield of the resources they exploit. Solving these problems demands changes in incentives, so that people respond appropriately to true costs and opportunities.

The incentive problems (Ferguson-Bisson, 1992) have the effect of turning rural people into 'disinterested land managers'. They include all measures which prevent people from being certain as to the ownership of the land and of its eventual output. Their primary interest remains that of exploiting the land to the extent possible now, as the right to do so may be lost any time. The dynamics is the same whether the lands concerned are cultivated, grazing or forest lands. Where lack of a perceived stake in land's future has been shown to turn people into 'disinterested land managers', lack of access to education and technical knowledge creates 'ill-informed land managers.' As a consequence, although often unknowingly, both categories contribute to land degradation.

Numerous government policies not only fail to induce users to base their decisions upon the true opportunity cost of resource use, but they also encourage more rapid and extensive forest degradation than would market forces alone. Many current policies including subsidies, taxes and market interventions artificially increase the profitability of activities that result in serious resource degradation.

The distinction between market failure and market distortions is also made by Sharma and Rowe (1992). They argue that

> in many countries where the government is the principal holder of forest property rights, traditional systems of providing access to forests and allocating common property resources to local people have broken down. Often a government's disregard of traditional rights of local people makes forests more vulnerable to open-access problems. Moreover, in many instances, governments lack the capacity to manage forests effectively and

control access to forest land under public ownership. Public policies could help compensate for these failures, but to date, by distorting the real (or scarcity) cost of forest resources, they have done just the opposite. Indeed the experience in many developing countries shows that agricultural incentive policies, land tenure, colonization, resettlement, taxation, credit and trade policies have often encouraged the expansion of the agricultural frontier at the expense of forests. Further aggravating matters has been the severe underpricing of tropical timber through deficient royalty and concession policies, inefficient fuelwood policies, and the weak enforcement of existing regulations and concessions.

Agricultural policies

Governments intervene in many ways in agricultural markets. Agricultural pricing policies play an important role in encouraging resource degradation. Agricultural output prices in developing countries tend to be depressed, whereby the internal terms of trade are turned against agriculture (Repetto, 1989). Repetto argues that:

> in general, depressed agricultural prices lower the farmers' incentives to practice soil conservation.[4] Furthermore by heavily subsidizing inputs of chemicals, capital, and water, farming systems which diminish soil productivity and the self regulating capacity of agricultural ecosystems are promoted and the emergence of alternative more sustainable forms of agriculture are prohibited even when they would be more productive and stable in the long run.

A more elaborate discussion of agricultural input subsidies follows in Chapter 5 (pp.85–136).

Many countries actively encourage the conversion of tropical forests for agricultural purposes. Rules of land tenure in many states allow private parties to obtain title to forest lands by showing evidence of 'improving' it – by clearing away the trees, for example. In the Philippines, Brazil and elsewhere, recognized rights of occupancy or possession are awarded on the basis of the area of land cleared. Such provisions often become a mechanism for privatizing land from the public forest estate. Those who obtain ownership soon sell out to larger landholders, who consolidate the land to establish private ranches and accumulate speculative holdings. In many cases such activities would be uneconomic without heavy government subsidies.

More general agricultural policies contribute indirectly to deforestation (Repetto, 1989, 1990). In Latin America and the Philippines the aggregation of the better agricultural land into large, generally underutilized estates pushes the growing rural population into forested frontiers and upper watersheds. The extreme concentration of landholdings is supported by very low agricultural taxes that make farms and ranches attractive investments for people in

---

[4] The assumption of a positive relationship between agricultural output prices and soil conservation is not undisputed, however. For instance, Barret (1991), in his analysis of this relationship, concludes that price increases will have no effect on soil conservation if farmers control soil fertility by adjusting the length of the fallow period, or even that price increases induce farmers to plant crops that cause greater damage to the soil.

upper-income brackets, for whom it costs almost nothing to keep extensive holdings that generate relatively little income. Subsidized rural-credit programmes also promote land concentration: ceilings on interest rates inevitably lead banks to ration credit in favour of large landholders who have ample collateral and secure titles. Governments, especially in Latin America, have offered generous fiscal and financial support to the agricultural sector. These generous fiscal inducements have led ranchers to convert large forested areas to extensive pasture in areas such as the Amazon (see also pp.262–284). Under minimal management and without adequate fertilization, many converted soils lose nutrients through leaching and are invaded by weeds, so that within a few years productivity declines and the pasture is abandoned. Nonetheless, such operations may still be privately profitable because of the policy incentives to investors.

Forest policy and management

Most governments – which are globally the proprietors of at least 80 per cent of the mature closed-canopy tropical forest – have not put an adequate value on that resource:

> while they sacrifice enormous sums in potential forest revenues, governments in the tropics are failing to invest enough in stewardship and management of the forest. The stumpage value, or economic rent, of mature virgin tropical forest timber is substantial. Many governments have offered timber concessions to logging companies on terms that capture only a small fraction of this rent in royalties, taxes, fees, leaving most of it as above-normal profits for private interests.
>
> (Repetto, 1990)

Repetto (1989, p.81) gives some examples of government rent capture in tropical timber production as a percentage of the actual rent:

Table 1.3

Country	Period	Timber/actual rent (%)
Indonesia	1979–82	37.5
Philippines	1979–82	14
Sabah	1979	82.6

Source: Repetto, 1990.

Governments not only fail to capture a fair share of the economic rent of their timber resources but also fail in the control on prohibitions of logging and on legal constraints that are laid upon the concession holders (for a discussion of the conditions under which concessions are granted, see pp.49–55).

Distorted incentives also reduce the efficiency of wood-processing industries. Many countries, such as Indonesia (see Appendix 2) seek to increase both employment and the value added to forest products domestically by encouraging processing rather than the export of logs. They must strongly support local mills to overcome high rates of protection against the importation of processed wood in Japan and

Europe. However, extreme measures, such as bans on log exports or export quotas based on the volume of logs processed domestically, have created inefficient local industries.

Many countries are currently taking steps to capture resource rents at full value. A number of governments are now strengthening their forest management capability with the help of development-assistance agencies. The World Bank and the Asian Development Bank now have loans for forest management improvement in the pipeline for a dozen countries. Most of these loans support forest-policy reform as well as institutional strengthening. Under the aegis of the Tropical Forestry Action Plan more than 50 countries are preparing national action plans to conserve and manage their forests.

Other policies

In many countries, deforestation has provided a temporary escape valve – a respite from development pressures that can be dealt with effectively only at a more fundamental level.

Governments are reluctant to take fundamental steps, however, e.g. to address population control directly or to attack highly skewed patterns of landholding. Precisely to avoid such unpopular measures, governments invest in the colonization of marginal lands, e.g. the forests. Investments in low potential regions yield relatively low returns; in many cases, the benefits of such investments do not exceed the economic costs. Because the costs of bringing low capacity, or 'marginal', lands into production can be very high, it is economically preferable for local governments to invest in measures to reduce population growth, thereby minimizing the need to make expensive investments in expanding the carrying capacity of marginal lands. The experiences of Indonesia's Transmigration Program illustrate this point. According to Mahar (1985), the Indonesian example illustrates that investments in family planning programmes to reduce population pressures in high-potential regions would be more effective as investments in expanding the carrying capacity of low-potential regions. Other examples of economically and environmentally costly land settlement schemes, stimulated at least in part by population pressures in high-potential areas, include the Northwest Region and Transamazonica Highway projects in Brazil, several settlement projects in the Amazon regions of Peru and Colombia, and Kenya's Bura Irrigation Settlement Project.

Indeed, land extensification has been a deliberate national policy in some countries where it has been seen as an alternative to land reform for relieving potentially revolutionary pressures upon land in traditionally densely settled areas (e.g. Northeast Brazil and highlands areas of Colombia, Ecuador, Bolivia, and Guatemala). In both Bolivia and Brazil, however, the new frontier settlements became over-populated soon after they were opened up because of the huge number of landless and near-landless people who poured in, searching for land as spontaneous migrants, thus further accelerating deforestation rates in the areas made accessible for colonization schemes.

2

FOREST PRODUCTS

INTRODUCTION

Although the conversion of forests into agricultural land is by far the most important cause of deforestation in the humid tropics, unsustainable forestry practices also contribute significantly to deforestation rates, as came to the fore in the decomposition of deforestation processes (pp.13–17). Tropical rainforests are exploited for a number of products, from three main product groups:

1) **Industrial wood**, for domestic use and export. Harvesting practices of commercial loggers usually lead to degradation or even destruction of the forests they have logged. Despite the destructive methods employed, commercial logging is often encouraged by local governments, because the export of logs and sawnwood is a source of foreign exchange receipts. Furthermore there is the interlinkage with deforestation by shifting cultivators: when forests are opened up for harvesting, shifting cultivators have better access to it, which may contribute to deforestation processes.

2) **Firewood and charcoal**, used locally or regionally. Over two billion people use wood and/or charcoal to cook their daily meal. Deforestation and lack of firewood have led to the so-called 'poor man's energy crisis'.

3) **Non-wood-products**, like rubber, bamboo, resins, spices and flowers. These products can be harvested on a regular basis without damage to the ecosystem. Exploitation of the forests for these products is therefore often regarded as a sustainable alternative to logging. But, as past examples show (Ryan, 1991) the popularity of some of these products led to overexploitation, resulting in environmental degradation.

The contribution of forest exploitation to deforestation varies per region. Commercial logging, for instance, is of more importance in Southeast Asia (Indonesia, Malaysia (Sarawak)) and Africa than in Latin America. In Africa, the search for firewood is a more important cause of deforestation than in the other two regions. In the following sections the domestic importance of non-wood forest products and their potential for sustainable forest management will be discussed. On pp.35–47 a description of the international tropical timber market, the theory of the use of trade policy instruments, and an evaluation of trade policy proposals, will be presented.

THE DOMESTIC TIMBER MARKETS

The major share of tropical hardwood products is used domestically. In 1987, for example, only about 27 per cent of total industrial roundwood production by tropical regions was exported (see Table 2.4, p.36). The importance of domestic consumption to the commercial forestry sector, however, varies considerably between countries. For example, in Indonesia in 1989, 20m³ of a total production of 39m³ of industrial wood was used domestically, whereas Nigeria and Brazil, two other large producers with a large population, direct most local production to the domestic market (Bourke, 1991).

On the whole, domestic use in developing countries has expanded significantly in recent years. Total apparent consumption in the developing countries rose from 161m³ in 1970 to 395m³ in 1989 (Bourke, 1991). Much of this increasing importance of domestic demand corresponded to a parallel growth in the size of the population. Moreover, as the projections in Tables 2.1 and 2.2 suggest, the consumption of the main forest products in the developing countries will continue to grow strongly. According to FAO projections (1986, in Regeringsstandpunt, 1991, p.106), the total use of industrial round-wood (hard- and softwood) will rise by 3.5 per cent annually till 2000. This means a 58 per cent increase from 1987 till 2000.

This anticipated growing domestic demand for industrial roundwood will partly be caused by population growth and partly by growing welfare levels. Comparing the demand for industrial wood per capita of developed countries with that of developing countries (1.1 m3 and 0.09 m3 respectively – World Watch Paper 83 and SBH, data 1986 in Otto, 1990, Table 2), a higher welfare level in the developing countries will probably result in a growing demand. If the welfare level of the developing countries would rise to the same level as that of the western countries currently, and the demand for industrial wood rose accordingly, the use of industrial roundwood would rise from 360 million to 4100 million cubic metres a year!

According to the World Watch Institute (Otto, 1990, p.4), about 130 million hectares of deforested land must be reforested in the next ten years to meet the augmenting demand. That is over ten times more than what is planned at the moment. During an international conference for environment ministers in November 1989 in Noordwijkerhout in the Netherlands it was agreed that efforts should be made to expand the

Table 2.1 Projected average annual growth in consumption of forest products, 1986–2000 (%)

Products	Developed countries	Developing countries
Sawnwood	2.0	3.6
Wood-based panels	4.4	6.3
Paper and paperboard	3.5	4.8

Source: Bourke (1991)

Table 2.2 Developing country share in world consumption of forest products (%)

Products	1986	2000*
Sawnwood	22	30
Wood-based panels	14	19
Paper and paperboard	16	24

* Projections.

Source: Bourke (1991).

forest areas replanted worldwide by 12 million hectares annually by the beginning of the next century (BOS, August 1991, p.33). Other ways to meet the increasing (domestic) demand for timber are to expand plantations or to expand the number of indigenous species that are accepted by markets. The latter options require that appropriate processing facilities are developed, since plantations, as well as new species, have very different characteristics from many species currently being used. The number, quantity and amount of timber products supplied to the domestic markets could also be expanded by improving processing technology and marketing expertise. In many countries there is room for improvement because domestic processing facilities, especially sawmills, are antiquated, inefficient and unable to produce acceptable products profitably.

According to Bourke (1991), the consumption patterns between the developed and developing countries differ in that domestic consumption in the latter is heavily skewed toward relatively unprocessed products, and sawnwood in particular. Of major importance to the level of domestic consumption and the uses of wood are a country's level of development, the sophistication of its processing sector as well as its access to materials, and the degree of its market orientation (Bourke, 1991). Product availability and therefore the consumption of industrial roundwood in many developing countries is generally linked to exporting activities. In most cases, the main domestic end-uses are in construction, infrastructure and industry (mining, railway sleepers, transmission poles, etc.), packaging and furniture. Bourke further argues that:

> the quality of the material used and of the products produced is generally low. Little sawn timber is graded or seasoned and few countries have sound and enforced quality-control systems to ensure that the material is adequate for many end-uses. Two main factors influence this. First, the financial ability of consumers is limited and they therefore look for the lowest price. As a result even larger sawmills, which could provide graded, dried and seasoned timber rarely do so since the buyer is unwilling to pay the additional cost required. Secondly, where demand exceeds supply or monopoly conditions exist, as is the case in some developing countries, there is little incentive for the producers to improve their product since they are able to sell all of their output if prices are kept low. A further factor contributing to low-quality raw materials and products derived from them is the fact that domestically orientated producers tend to be small

and dispersed, with relatively unsophisticated or poor equipment and limited market information.

Another major barrier to improved processing and marketing of wood in tropical wood producing countries, both for the domestic and export markets, is transport. The transportation of logs as well as the processed products is often a difficult and time-consuming process. Logs must often be transported over long distances from isolated forests to the ports or the major population centres, where processing plants usually are established. From these locations products are transported to other centres, often by individuals who then sell to other entrepreneurs or small manufacturers.

A further constraint to the effective operation of domestic timber markets is the often excessive government influence. Local governments influence the domestic markets for industrial roundwood in many ways. At the supply side, the government usually controls much of the forest resource and, in many cases, a significant share of the logging, processing and distribution of forest products. At the demand side, government infrastructure programmes, government building programmes and activities that need concrete forming or pallets for transport utilize a large part of the products produced. At the policy level, governments influence consumption through a variety of processes, including housing loan policies, public housing schemes, rent controls, and import restrictions. Further factors which inhibit an effective operation of the domestic markets are inadequate market information and inadequate training of management and staff.

Another role which governments play in deforestation is the lack of control on prohibitions of logging and legal constraints laid upon the concessionholders. The concessionholders are therefore not stimulated to produce sustainably, for instance because of too short lease periods. A more elaborate discussion on the conditions under which concessions are granted will follow in Chapter 3. Governments that shelter inefficient processing industries can incur heavy economic and fiscal losses. Finally, governments tend to underestimate the gains of the non-wood products and of the services, like ecotourism, that can be derived from the primary rainforests.

FUELWOOD CONSUMPTION

Biomass fuels – mainly firewood – still account for 35 per cent of energy supplies in developing countries (World Bank, 1992). In Africa, for example, around 90 per cent of the population use fuelwood for cooking, the equivalent amount of roughly 1.5 tons of oil per family per year. Biomass is used not only for cooking but also in small-scale service industries, agricultural processing, and the manufacture of bricks, tiles, cement and fertilizers. Such demands can be substantial, especially in and around towns and cities. Fuelwood consumption in developing countries, especially in Africa, is increasingly leading to deforestation because the demand for fuelwood is being met increasingly by the consumption of tree stocks. The depletion of existing stocks further aggravates the problem because where consumption is increasing exponentially with population growth, the self-renewing capacity – the annual addition to supply through new tree growth – is declining in proportion to the volume of stocks (Anderson and Fishwick, 1984; Myers, 1992). In Table 2.3, the annual 1965-89 fuelwood consumption for the world as a whole and for some principal regions, together with the average annual 1965-1980 and 1980-1989 growth rates, are presented.

In 1989 sub-Saharan Africa accounted for 23 per cent of total world consumption of fuelwood, a share only surpassed by East Asia and the Pacific (24 per cent), but growing at a far faster rate (3.2 per cent against 1.9 per cent for East Asia and the Pacific). If wood fuels remain the staple source of household energy in sub-Saharan Africa with 90 per cent of households using them for cooking, the likely high sub-Saharan African population growth rates (see also ch.6) will cause fuelwood consumption in the region to increase rapidly, for a rise in population translates directly into an increase in demand for wood fuels. Many regions in sub-Saharan Africa are already confronted with fuelwood shortages. These shortages have three primary causes:

1) An increase in wood fuel consumption;
2) Expansion of agriculture into forests and woodlands which reduces available tree stocks; and
3) Overgrazing caused by an increase in the cattle population.

(Barnes, 1990).

Except for a limited area protected as forest reserve (see Table 3.1, p.57), in most African countries the exploitation of the remaining forests is virtually unmanaged, even if the sales of firewood and

Table 2.3 Fuelwood and charcoal consumption
(millions of tons of oil equivalent)

	1965	1970	1975	1980	1985	1989	Average annual growth rate (%) 1965–1980	1980–1989
World	244	263	286	329	372	399	1.8	2.2
Low- and middle–income	198	228	254	281	315	343	2.2	2.3
Sub-Saharan Africa	46	53	61	71	84	95	2.9	3.2
East Asia and Pacific	60	67	75	83	91	98	2.1	1.9
South Asia	44	50	56	62	70	76	2.3	2.3
Latin America and Caribbean	37	41	46	52	58	62	2.4	2.1
Europe	9	13	12	8	7	7	-2.8	-1.7

Source: World Bank (1992, Table A.10, p.205).

charcoal for urban and industrial use represent a multimillion dollar business that employs tens of thousands of people (Armitage and Schramm, 1989). Forests are commonly exploited on an unsustainable basis because of incentive problems arising from market failure. In the case of the markets for fuelwood the following problems persist (Armitage and Schramm, 1989):

- 'the economic value of the standing stock of wood cannot easily be recovered from those who cut it down. In many countries, neither the land on which the wood is standing nor the wood itself belongs to those who cut it. Therefore, there is no incentive for long-term management that would lead to sustainable production of wood, nor is there an incentive to maximize yields. Even though wood cutting and charcoal making are often illegal, such activities are practically never policed', and;
- 'there is no incentive for replenishing the dwindling supplies of wood through private replanting, because prevailing market prices reflect only the costs of production from standing stocks, but not the additional high costs of replacement.'

There are three reasons why market prices do not reflect the real value of fuelwood, given its growing scarcity:

1) The problem of environmental discontinuities (Myers, 1992) or the stock-flow problem, which causes scarcity to be revealed only when the depletion process is already long under way. When cutting starts to exceed regrowth, stocks necessarily must decline, but this process is slow and imperceptible at first. Only after a certain threshold, where the ratio of standing stocks to net incremental growth is over seventy, do stocks suddenly decline sharply.

2) Market prices are kept low and only reflect cutting, production, and distribution costs because of the severe competition in the commercial firewood and charcoal business, combined with the fact that the resource base is common property.

3) Because in many countries the wood for charcoal making comes mainly from agricultural land clearing operations, it has no alternative value.

Policy options (Anderson and Fishwick, 1984) which would relieve pressures on the declining resource base include:

- **Substitution** – principally by commercial fuels (kerosene, LPG, electricity) for fuelwood and charcoal in urban areas;
- **Conservation** – through the encouragement of changes in people's cooking methods in ways that save fuelwood; and through increased efficiency in harvesting (reduction of wastes) and in charcoal conversion methods; and
- **Increasing supplies** – through investment in forestry and agroforestry.

It is highly unlikely that a massive transition from biomass fuels to commercial energy sources will occur in the near future. The most important reason why substitution (or any other adjustment) stagnates is probably the treatment of the forests as a public good (Anderson and Fishwick, 1984). As a consequence the degradation of these forests is not included in the price of fuelwood, resulting in a fuelwood demand which is higher and a rate of substitution which is lower than is economically desirable. As long as this situation prevails, autonomous demand adjustments will lag behind what is required to conserve the forests.

In any case, as past evidence suggests, autonomous transition in urban areas in the face of rising scarcity will probably only occur if forest resources are already badly depleted, usually over a radius of 50-100 miles or more. Wood fuels are preferred even where the real costs of cooking with fuelwood have risen above the costs of cooking with commercial fuels, as is the case in a large number of African towns and cities (Anderson and Fishwick, 1984). Indeed, in many urban areas there is an economic gain from substitution, even ignoring the ecological benefits of reduced fuelwood consumption. The obvious recommended policy is therefore to encourage substitution and energy conservation in urban areas, including the regional towns and cities. But even in countries where kerosene is subsidized to encourage substitution the results are meagre (World Bank, 1992): 'this leads to some extra substitution, but people buy excess amounts and retail it as a (very polluting) substitute for diesel fuels'. In rural areas, instead, there is usually no economic gain from substitution because of the high costs of distributing commercial fuels, given the distances involved and the poor infrastructure.

From a local government point of view, a continued use of domestic woodfuels has several advantages. An important advantage for countries where foreign exchange is scarce - which is the case in most of the tropical countries - is that domestic woodfuels as compared to imported fuels, have a relatively small impact on foreign resources (limited largely to the costs of transportation through imported petroleum fuels for trucks and foreign spare parts). The rate of substitution between fuelwood and other energy consumption therefore depends on a country's capacity to finance imports, that is on the success of their trade and exchange rate policies. The overvaluation of currencies in many African countries, for example, combined with the relative neglect of the agricultural sector as compared to industry, has been discouraging to private investments in rural areas, including investments in more sustainable production methods such as agroforestry (Anderson and Fishwick, 1984). Another important 'advantage' of the use of domestic fuels as compared to imported fuels is that fuelwood cutting and charcoal production provide urgently needed additional employment for a large number of unskilled and semiskilled labourers during low agricultural activity periods.

Fuelwood conservation

The introduction of woodburning stoves, and policies to change people's habits and methods of cooking, is widely regarded as a promising way to save on fuelwood. Experiments have shown that a significant improvement of fuel efficiency can be obtained:

Two- or threefold increases in the fuel efficiency of cooking would be possible over traditional 'three-stone' methods. In addition, the use of aluminium pots, whether on open fires or on stoves, has been found to add significantly to efficiency, with further gains being thought possible through improvements in cooking habits and utensils. Overall it was once thought that a fourfold reduction in the specific consumption rates of fuelwood could be achieved by these various devices, from around 0.8 cubic meters per capita per year, to 0.2 cubic meters.

(Anderson and Fishwick, 1984)

Regarding the introduction of improved biomass stoves, the World Bank, for example, states that: 'The potential benefits of stoves programmes are considerable. In addition to the large direct benefits of fuel savings, recent research has found that the economic value of the environmental and health benefits of improved stoves amount to $25–100 a year per stove, leading to a payback period to society of only a few months.' (World Bank, 1992, p.127).

It is obvious that a large-scale transition to such fuel conserving methods could greatly reduce deforestation. Policies to improve fuelwood efficiency are even more attractive because the costs of these programmes are relatively small, that is, only a very small fraction of, for example, net public investments in energy (Anderson and Fishwick, 1984). An improved biomass stove, for instance, can be made locally by artisans, with payback periods of 20–60 days assuming the efficiency gains are realized (World Bank, 1992).

So far, however, programmes to put these ideas into practice have met with mixed success. Programmes which were successful have a number of characteristics in common, such as (World Bank, 1992):

• they have concentrated on the users most likely to benefit. The people who first adopt improved biomass stoves are usually not the poorest but those who have limited income and are spending much of it on cooking fuel;
• the designers and producers of the stoves discuss them with each other and with the users;
• the program relies on mass-produced stoves and stove parts, which seem to be more successful than custom-built stoves; and
• subsidies go towards the development of stoves rather than to consumers for purchase of the stoves.

Increasing supplies by replenishing tree stocks

Where substitution policies and fuelwood conservation programmes are addressing the fuelwood scarcity problem through adjustments of fuelwood demand, policies to encourage investments in forestry and agroforestry are aimed at adjustments of the supply of fuelwood. Forestry plantations and fuelwood plantings currently account for a very small percentage of supplies. In Africa, for example, more than

95 per cent of total fuelwood supply stems from naturally grown forests and woodlots (Anderson and Fishwick, 1984). To meet projected future fuelwood demand without further serious degradation of the remaining forests, however, extending the existing private fuelwood plantations, and establishing new ones, seems imperative.

The main difference between the two (complementary) approaches is that the forestry approach is aimed at an increase of public fuelwood resources, and the agroforestry approach is aimed at the increase of private woodlots. The forestry approach involves the extension of plantations, the better management of forest reserves, the recovery of forest residues, and the establishment of fuelwood and erosion control plantations in watersheds and shelterbelts. This approach works through the forestry services. The agroforestry approach (as discussed on pp.122-136), involves the planting and maintenance of trees near dwellings, on the boundaries of arable land and in small lots (sometimes intercropped) by the farm families themselves.

Past evidence in Africa suggests that increasing supplies through the forestry approach is a promising strategy to address the fuelwood problem. In Africa the larger share of fuelwood related investments by governments and donors have been allocated to forestry projects. These projects show that with rising scarcity prices, fuelwood plantations have 'prospects of good financial and economic rates of returns, e.g. as with peri-urban plantations for towns and cities where local reserves are badly depleted, and multi-purpose plantations providing fuelwood, poles and erosion control in rural areas' (Anderson and Fishwick, 1984).

Although a promising strategy, the possible increase of future fuelwood supplies by the forestry approach will not be sufficient to meet projected demand. The expectations about the increase of fuelwood demand are such that the contribution of forestry related policies will be limited as a consequence of limits to financial and institutional capacities. Anderson and Fishwick (1984) for instance, state that 'even if demand for wood were to be reduced by 20 to 30 per cent through conservation, and through substitution for other fuels, a fifteenfold increase in current planting rates would be required in African countries to bring demand and supply into better balance by the year 2000'. It is because of these vast requirements that the forestry approach is described as a complementary solution to the agroforestry response to the fuelwood supply problem, rather than a solution on its own.

With respect to its potential to increase tree stocks, the agroforestry approach, as compared to other policies, has distinct advantages. First, farmers often outnumber foresters by 1000 to 10,000 to 1. If their families spend, say, only ten days per year on the planting and care of trees, as an off-farm activity, their combined labour inputs would outweigh that of the foresters by 40 to 400 to 1, and secondly, only small areas of farmlands need to be taken up to be practical. For Northern Nigeria it is estimated by Anderson and Fishwick (1984) that 40-50 trees planted on farm boundaries and near dwellings, occupying less than 5 per cent of arable land on, say, a two hectare farm, is sufficient to meet the fuelwood needs of a farm family. One of the reasons is that the yields of free-standing trees are typically three or more times those in plantations and forests.

Another advantage of agroforestry as compared to the forestry approach is that the unit costs of programmes to encourage agroforestry practices are very low – of the order of 10–20 per cent of the costs of plantation projects, possibly less. The transition to agroforestry as a means to increase fuelwood supply, however, just as in the substitution of other fuels, will stagnate as long as local tree stocks can be cut free of charge. As long as farmers can obtain their fuelwood from local forests without being charged for the replacement costs of the trees they cut, and the maintenance costs of replanted stocks, these farmers have little incentive for private plantings. The transition to agroforestry systems will furthermore lag behind what is required because the ecological costs of overcutting are indirect and not immediately apparent.

The theoretical answer to the incentive problem for private tree planting, as well as for conservation and for substitution, is to increase extraction costs with a stumpage tax. The level of this tax should be such that the costs of planting new stocks and tending them to maturity is recovered. The implementation of such a tax could provide funding for fuelwood demand or supply adjustment programmes. In practice, however, the administrative and political difficulties of applying such a tax over large areas, and of protecting the forest reserves, are formidable. In cases where stumpage fees are applied, mainly for timber concessions, these fees are often set far below the replacement costs (see Chapter 3).

OTHER FOREST PRODUCTS

Recently non-wood products originating from the tropical forests have started to receive much attention. It is widely believed that forests could be commercially exploited for these products on a sustainable basis. A wide and ever-increasing range of non-wood products varying from legumes, fruits, juices, nuts, mushrooms, fish and meat to livestock fodder, oils and fats, gums, fibres, medicines and raw materials for the pharmaceutical industry, stem from the forests. Tens of millions of people are dependent on the tropical forests for their food, health care, raw materials and cash income. Most of these products, however, are used for subsistence or traded locally. Their economic value is therefore hard to monitor and easily overlooked. Indeed, where timber is a high-profile export product and therefore highly visible, few statistics are kept on the economic value of other forest products, products which if they do appear in official statistics are ranked under the heading 'minor'.

Although the economic feasibility of the industry based on these products is not as visible as that of the large-scale wood based industries, it may be as great or even greater. This can be illustrated by examples of the economic importance of those forest products which are internationally traded. In several cases forest-based industries have become important export earners. In Southeast Asia, for instance, rattan is a major income source for many countries; the value of total rattan trade is estimated at more than $3 billion a year (Ryan, 1991); with export earnings for Indonesia of $90 million, for the Philippines of $90 million, for Malaysia of $35 million, and for Thailand of $15 million (Asian Development Bank, 1989). Another important income source for Southeast Asian countries is bamboo, a material which has long supported a great deal of processing both for domestic and foreign markets. South American examples, besides cocaine revenues, are the exports of chicle, allspice and xate which earn Guatemala $7 million annually; and the 1989 Brazilian exports of palm heart and Brazil nuts with an estimated value of about $20 million each (Ryan, 1991). An example in Africa is the export of honey from forest bees in Tanzania, which is several times more valuable than the timber produced in the nation's forests (ibid.).

Where timber production is usually a capital intensive business, using heavy machinery and relatively few people, with a non-local, often foreign management, non-timber forest products instead are harvested by local people using simple tools only. Many people earn a living in forest-based industries such as in the production of furniture, baskets, luggage, mats and a multitude of handicrafts. Other important forest-based activities are processing of wildlife products for hides and skins, gums, essential oils and dyes, the production of mushrooms and the collection and processing of medicinal supplies of forest origin. As a result, in some cases timber does not employ as many people as alternative forest products. For example, Indonesia's rattan industry supports more than 200,000 full-time employees, three times that of its timber industry (Asian Development Bank, 1989).

Because non-timber forest products form an undervalued part of the economies of tropical countries, they are undervalued in economic decision-making as well. Most governments still regard forests primarily as stocks of timber or as potential farmland, without sufficient regard for other opportunities. However, recently an increasing number of studies demonstrate the underestimation of the economic value of other forest products or even indicate that they often are worth more than the timber in the research area. In a report on the economic value of non-timber forest products in Southeast Asia, De Beer and McDermott (1989), for example, conclude that the value of non-timber forest products (rubber, fruit, nuts etc.) which can be exploited by local communities is usually significantly underestimated. In an article on the value of non-timber resources of the Amazonian rainforests, Peters et al. (1989) demonstrate that non-timber resources in one hectare of Amazonian jungle are worth more than the timber itself and therefore conclude that the sustainable exploitation of non-wood forest resources represents the most immediate and profitable method for integrating the use and conservation of Amazonian forests.

An important advantage of the exploitation of forests for non-timber products is that most of these products can be harvested far more regularly than timber. The annual net revenue of other forest products (edible fruits, cocoa and rubber) from one hectare of Peru's Amazon forest, for example, was estimated to be, over time, $6330 per ha. This is approximately six times the amount that could be earned from harvesting all the timber in a single year or twice the value of converting the land to cattle pasture (WRI, 1991). It should be noted, however, that this study covers only a small fraction of the forest, containing relatively large quantities of the marketable products, which serves an entire regional market. The question therefore remains what the market value of these products would be if larger areas could be harvested and the quantity supplied increased.

The various studies on the economic value of non-timber forest products generally tend to support the idea that it could be profitable to save the tropical forests. Such an alternative sustainable use of the forests would also secure the livelihoods of the people dependent on them. An increased awareness of the value of the forests for their non-timber resources could dissuade governments and individuals from destructive uses such as logging and cattle ranching. The joint effect of saving the forests and its people therefore motivates non-profit groups and green-minded business, encouraged by environmentalists, to introduce all kinds of forest products in European and American markets.

One should not be too optimistic about this option, however. Indeed, it is questionable whether, if exploitation of a forest intensifies as a consequence of an increase in the commercial value of its products, its ecosystem remains intact, for the usual fate of species that gain long-term popularity in industrial markets is depletion. In Ryan (1991) some examples of this form of forest degradation are given, such as in the lowland jungles of Peru where, although there has been little deforestation, environmental degradation instead has resulted from the depletion of forest species that have become popular with urban consumers. In this region, as in Brazil, entire palms are destroyed to harvest their 'hearts' for export. As a consequence of overfishing and overhunting virtually all large fish and wildlife species are also in decline. In Southeast Asia, the increased demand for rattan and other forest products combined with high deforestation rates, has resulted in serious forest degradation.

As for the situation of the local producers, Ryan argues that: 'the usual lot of extractors is continued poverty, as the profits from their work are siphoned off by powerful middlemen and elites. Brazil-nut gatherers, for example, receive about four cents a pound for their labours, just 2 to 3 per cent of the New York wholesale price. Three fourths of the Brazil-nut market is controlled by three companies, owned by three cousins.' (Ryan, 1991)

The conditions for many of the rubber tappers in the Brazilian Amazon are more or less as miserable: 'most are chronically in debt and exploited by middlemen. As the murders of Chico Mendes and perhaps

1500 other rural people in Brazil's interior during the past decade attest, landowners continue to use violence against anyone who threatens their economic dominance' (ibid).

In cases of overexploitation the situation of local producers is even worse, as they have to work longer or receive less income, or both.

A promising strategy to cope with these developmental and environmental problems could be the establishment of extractive reserves (World Bank, 1992; Ryan, 1991; WRI, 1990)[1]. These are tracts of forests set aside as a source of natural products that can be harvested indefinitely without clearing the land. The approach differs from other forms of protection in that the reserves are managed by local communities, which have legal rights to collect the products from their traditional territories. These rights form an incentive to the local managing communities for a sustainable use of the forests and to protect these forests from encroachment. The first extractive reserve was established in the Brazilian Amazon, where the government, pressed by the rubber tappers union, granted secure titles to the use of the forests it contained. Titles restrict the right to extract to individual families of territories within the reserves. Individual property rights are not granted. The land cannot be sold or converted to non-forest uses; only some small scale subsistence farming is permitted. Up till now 14 extractive reserves covering a total of 7.5 million acres in four states have been created by the Brazilian government for the use of the rubber tappers living in them. The first six of these reserves were established in Acre, one of the states most threatened by deforestation (World Bank, 1992). Twenty reserves, covering in total 250 million acres, have been proposed. If these proposals are accepted it would mean that about 25 per cent of the Brazilian Amazon will be set aside as extractive reserve.

Although promising, this does not mean that the establishment of extractive reserves alone is sufficient to protect the tropical forests (Ryan, 1991). Firstly, although extractive reserves are protected from clearing, fully undisturbed reserves are still crucial for preserving all the environmental values of a tropical forest ecosystem. Secondly, the revenues of non-timber products alone cannot provide for what is required by the rural population. Furthermore, forest extraction is usually interlinked with shifting cultivation: most forest gatherers also clear patches of forests for farming, and most small farmers depend on nearby forests for additional food and other products. In this respect according to Ryan (1991) 'because they can only benefit a fraction of the growing number of people that are forced to destroy forests to survive, they cannot be a substitute for programmes to redistribute farmland, reform farming practices, and halt population growth.'

---

[1] For a more elaborate discussion of strategies for the management of timber and non-timber forest products see Chapter 3.

THE INTERNATIONAL TROPICAL TIMBER TRADE[2]

Trends in production and trade

During the past three decennia the production of tropical timber increased strongly in all stages of processing. The production of roundwood in the tropical countries increased nearly fourfold during 1960-1987, to a level of 150 million cubic metres in the latter year (see Table 2.4). The same trend applied for the production of processed tropical hardwood, mainly in the form of sawnwood, plywood and veneer.

As for the exports of tropical timber the picture is somewhat different. During 1960-1980 the exports of tropical roundwood increased to about three-and-a-half times the original level; however, thereafter it declined, during 1980-1987 from 38 to 25 million cubic metres. The latter development can be explained, firstly by the policies of major exporters - through restrictions on the exports of logs and through export levies - to encourage domestic processing,[3] and secondly by a strongly increasing domestic consumption. The exports by the tropical countries of processed tropical hardwood products instead rose strongly during 1960-1986.

At present Malaysia and Indonesia are the main exporters of tropical hardwood, Malaysia being the world's major exporter of tropical hardwood logs and veneer, and Indonesia the largest exporter of plywood (see Appendix 3). Both countries also dominate the international market for tropical sawnwood.

Figure 2.1 represents the total volume of 1988 world exports of tropical timber per principal region of destination, and clarifies that an important share of total timber exports relates to trade within the group of developing countries; the major part of the remaining share goes to Japan and the EU.

Table 2.5. subsequently focuses on the volume data of the regional distribution of hardwood forest products' international trade flows to

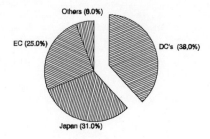

Figure 2.1 Exports of tropical timber per region of destination, 1988

---

[2] For a further analysis of this topic, see also De Bruyn, 1992.

[3] According to Repetto (1988, p.34) domestic wood processing industries supported commonly are inefficient in comparison to the competing western industries, insofar as they need more wood to produce the same amount of final product. The quality of their products seems to gradually improve; moreover their prices generally are low.

Table 2.4 Production and exports of the major hardwood products by tropical regions (million m³)

	Production				Export			
	1960	*1970*	*1980*	*1987*	*1960*	*1970*	*1980*	*1987*
Tr. America								
Roundwood	11.5	15.3	28.0	44.6	0.3	0.4	0.1	0.0
Sawnwood	5.1	7.1	12.6	19.5	0.1	0.5	1.1	0.9
Plywood	0.2	0.7	1.4	2.1		0.1	0.2	0.3
Veneer		0.1	0.4	0.4		0.1	0.1	0.1
Tr. Africa								
Roundwood	9.2	14.7	17.5	15.8	4.4	6.8	6.1	3.6
Sawnwood	1.5	2.6	5.1	5.7	0.6	0.7	0.7	0.8
Plywood	0.1	0.2	0.4	0.5	0.1	0.1		0.1
Veneer		0.2	0.4	0.5		0.2	0.2	0.2
Tr. Asia								
Roundwood	22.7	52.2	83.6	89.2	6.1	29.0	32.1	23.4
Sawnwood	6.5	11.8	24.9	32.9	1.2	2.5	6.4	8.6
Plywood	0.4	1.3	2.5	8.8	0.1	0.5	1.7	7.9
Veneer		0.4	0.6	0.7	0.1	0.2	0.2	0.6
Tr. regions total								
Roundwood	43.4	82.2	129.1	149.6	10.8	36.2	38.3	27.0
Sawnwood	13.1	21.5	42.6	58.1	1.9	3.7	8.2	10.3
Plywood	0.7	2.2	4.3	11.3	0.2	0.7	1.9	8.3
Veneer		0.7	1.4	1.5	0.1	0.5	0.5	0.9
Total	114.4	213.4	354.4	441.2	26.0	77.2	97.8	93.0

Source: FAO/ECE, 1986; FAO, 1989.

Table 2.5 Estimates of imports of major hardwood forest products from tropical countries by Japan, Europe and the United States: 1960, 1970, 1980, 1987 (million m³)

	1960	*1970*	*1980*	*1987*
Japan				
Roundwood	4.8	19.9	19.2	14.2
Sawnwood	–	0.3	0.6	0.9
Plywood	–	0.5	0.1	1.6
Veneer	–	0.1	0.1	0.2
United States				
Roundwood	0.3	0.1	–	–
Sawnwood	0.4	0.4	0.5	0.7
Plywood	0.5	0.2	1.0	1.5
Veneer	0.2	0.2	0.1	0.1
Europe				
Roundwood	3.6	6.4	5.9	3.4
Sawnwood	1.0	1.5	3.2	4.1
Plywood	0.1	0.2	0.7	1.0
Veneer	0.1	0.1	0.2	1.0

Source: ECE/FAO, 1986, and FAO, 1989.

the industrialized countries during 1960-1987. It shows again that the major part of the volume traded still goes to Japan, and also that the imports of the USA have remained fairly modest. The data also show, however, that Japan traditionally imports a relatively large amount of (unprocessed) roundwood logs for its wood-processing industry, whereas Europe and the USA instead import relatively more processed varieties, sawnwood and plywood in particular. This difference in the composition of the imports of timber products also explains why the value data differ quite significantly from the above volume data. Trade statistics on 1988 showed that the EU accounted for 43 per cent of the above import value, against Japan – responsible for 58 per cent in volume terms – that accounted for only 39 per cent of the import value (USA: 4 per cent).

The total value of the slightly less than 30 million cubic metres of tropical hardwood products traded worldwide by the end of the 1980s was some $8 billion. The total trade volume also reflects some patterns of regional specialization. More detailed data reveal for instance that EU imports of roundwood and veneer are mainly from Africa, whereas the major part of sawn- and plywood is imported from Asia.

The question arises what can be said about medium-term global trends with respect to exante supply of, and domestic and foreign demand for, tropical timber, and consequently about trends in international timber trade. Although much of the answer to this question will depend on such factors as the search for alternative resources, the finding of new usable species, the distribution of technical progress with respect to timber processing, the overall economic development, population pressure, the policy environment and policy measures, etc., many projections point in the following direction:

- Domestic demand for tropical timber will most likely continue to grow, due to increasing population pressure (firewood demand) and higher welfare levels (sawnwood demand); this may increasingly crowd out the timber resources left for exports.
- The supply of tropical timber may become adversely affected by the current deforestation rates and the consequent strong decline in the tropical forest area. Countries such as the Philippines, Thailand and Nigeria have already ended their timber exports for that reason; thus further reducing the number of timber exporting countries. Some projections indicate that this number could, during the 1990s, come down from some 30 to less than 10! The policy to decrease export supply due to unacceptable deforestation levels has already started in Asia, and will be carried over to Africa and Latin America (see also Appendix 4).
- The trend of an increasing volume of timber being processed domestically is expected to be carried on during the 1990s. This trend will be fuelled by the increasing restrictions in the timber producing countries on log exports, the improving quality of their processed wood products, improved technical knowledge of timber processing along with their relatively low wage levels. As a

result the shares of sawnwood, veneer and plywood in their production and exports are likely to rise at the cost of the share of roundwood.

- Less easily available tropical timber resources in the near future, higher government taxation on logging activities and higher concession pricing may all contribute to a gradual increase in the real costs of logging in the medium term. However, since logging costs are only responsible for some 15 per cent of consumer prices for timber products on average, the impact of these cost increases on the real consumer prices will remain fairly modest, and will mainly hold for roundwood.

Some additional restrictions on the speed of real timber price increases in practice can probably be explained by the fact that rapidly rising timber prices would induce not only additional (unsustainable) supply from areas that have remained fairly untouched thus far, or from new species, but also the search for substitution possibilities through other materials (see also FAO, 1986, p.319). The World Bank (in NEI, 1989, p.14), for instance, expects the real price of Southeast Asian logs to rise until 2000 by 1.1 per cent per annum only.

Timber trade policy options

Whatever the timber products may look like, in any case a discussion has started in various policy circles on ways to apply trade policies to internationally traded timber products as a means to speed up the process of sustainable exploitation of the tropical forests and to halt the worldwide rapid deforestation. The basic issue then becomes what the implications of the various trade policy options can be in terms of altered patterns of supply and demand and the (international) distribution of income and wealth between producers and consumers. Some of the fundamental insights derived from trade theory may serve to clarify the issue. If one restricts the analysis to one particular market (the timber market), and if one, for convenience sake, assumes that import demand and export supply schedules are at a normal level and are linear, an international timber market equilibrium in the absence of any trade policy or of any other market restriction (there is full competition) could be represented by point 1 in Figure 2.2; p^* and q^* represent the equilibrium price and quantity, respectively.

If one now introduces a uniform (*ad valorem*) import-tariff, t, a wedge is driven between the price paid for a unit of timber by the importer and the price received for the same unit by the exporter. This wedge, which equals t per unit of timber, accrues to the importing countries' government, who can use it to increase domestic government expenditures, to reduce domestic taxes or for both. If the vertical distance t in Figure 2.2 represents the size of the import tariff per unit of timber, market equilibrium is only compatible with the new situation if the volume of trade is reduced to q^{**}, and if the export prices and the prices in the importing country have become p^{**} and $p^{**}+t$, respectively.

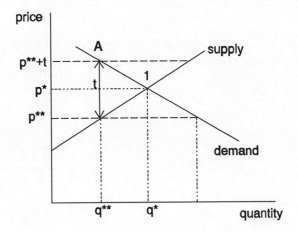

Figure 2.2 The effects of trade policy devices

The impact for the importing country will be that, although the domestic consumers seem to pay a higher price for timber, in fact they do not, if one takes into account that the government's tariff income will eventually somehow be transferred back to the same group of the domestic consumers. From a welfare point of view the country may well gain, especially if an import-competing domestic industry might take advantage of the reduced foreign competition. The exporting country, instead, will always stand to lose from the protectionism it has been confronted with, because it has to accept lower export prices, and thus a loss in its terms of trade. It speaks for itself that the terms of trade loss it faces will be larger if the importer imposing the import tariff has a sizeable impact on market conditions.

The point to be made now is that the precise impact on the market equilibrium of the given protectionist device, for instance applied to the tropical timber market equilibrium, very much depends on:

1) the kind of trade policy device that is actually applied;
2) the elasticities of supply, demand and substitution; and
3) the factors that may cause the supply and demand schedules themselves to shift.

One should therefore study carefully what the particular market circumstances are.

The various trade policy devices
If one somehow tries to influence the size and direction of the trade flows, one can either directly affect prices (and quantities indirectly), or one can directly try to affect quantities traded (and prices indirectly). In the first case, the market still functions albeit with biased prices; in the second case not. So, one would expect that it does matter how one tries to influence trade flows.

A second aspect with regard to the application of trade policy instruments is from whose side the measure is taken, that of the importer or of the exporter. Sometimes this aspect is irrelevant from the point of view of the international distribution of welfare; sometimes, however, it matters a lot.

If the two above aspects are combined one arrives at four policy devices (see Table 2.6).

Table 2.6 Trade policy options

	Measures taken at import side	Measures taken at export side
Prices directly affected	1 import tariff	2 export levy
Quantities directly affected	3 import quota	4 export quota

What difference does it make what measure will be taken? In theory all four measures can have precisely equal effects on trade flows, prices and the consequent welfare distribution. This however, is only the case if we accept a set of rather heroic assumptions. For the common effect of all policy devices is that they drive a wedge between the price the exporter receives and the price the importer pays for it. With respect to an import tariff this was illustrated already in Figure 2.2; if a quota is applied the same holds, because the demand curve then kinks downwards vertically at point A. So if the quota is binding, consumers are willing to pay $p^{**}+t$, whereas producers are satisfied with a price p^{**}, leaving a rent for those that succeed in getting the right to carry out the international trading transaction. The rent is t per unit traded, and therefore equal to the tariff that has the same trade impact.

So in all cases we are left with a margin in between. In the case of an import or an export tariff, this margin leads to tariff income for the government of the importing and exporting country, respectively. In the case of a quota the private sector will somehow capture the rent caused by this margin, unless the government decides to sell the 'right' to trade the commodity internationally, on the open market. With the theoretical case wherein the government would sell this right through an auction by tendering on the basis of international competitive bidding, one might expect that the full rent would be captured by the government, either of the exporting, or of the importing country, depending on the conditions of the quota. So, in extreme cases this would not differ from the outcome if a tariff were be levied.

If a tariff is levied by the importer (exporter), the importing (exporting) country's government will receive the tariff income. If the tariff revenues are not used for the benefit of the national citizens or taxpayers, however, but for the common cause – such as sustainable timber production, as far as timber trade is concerned – it is irrelevant which country's government acts as the collector of trade taxes, for in both cases the revenues will be transferred to the producers. The same would equally apply if the common cause was to somehow subsidize the consumers if they decided to only buy sustainably produced timber.

So, all in all, in theory the choice of the type of trade policy device can be irrelevant. In practice, of course, it matters quite a lot whether the rent is captured by the government or by the uncontrolled private sector, and also it will make quite a difference whether the tariff income (or the rent) will end up in the hands of the importing or of the exporting country's government.

Box 2 The Lerner symmetry theorem

According to Lerner's symmetry theorem, it does not make any difference to the allocation process if the tariff on international trade is levied by the importer (such as illustrated in Figure 2.2) or by the exporter. In case of an export tariff the case illustrated with the help of Figure 2.2 remains equally applicable: the *ex ante* supply and demand schedules remain the same, just as the size of t, at least if the export and import levy to be compared are the same value.

So, the impact on both, the volume of supply, demand, and trade, and on the various prices, will also be the same, irrespective of whether an import or export levy is imposed. The cases are fully symmetrical except for the government who will collect the tariff revenues. Depending on the tariff, this will be either the importing or exporting country's government. The above phenomenon equally applies if an import and export quota are compared, and is known as Lerner's symmetry theorem.

So, why is it important in practice where and how trade policy is applied?

Basically because the extreme conditions under which the impact of the various trade policy devices represented in Table 2.6 are the same, will seldom be met:

- Governments seldom sell quota rights at the market on the basis of competitive bidding, to capture the quota rents. Mostly they either simply distribute the quota among the importers without charging a price for these licences at all, or they leave the quota distribution to private initiative, which implies that the mechanism of allocation has become 'first come, first served' or – perhaps a better description if a sizeable quota rent is involved – 'survival of the fittest'.
- Even if a quota is imposed by the importing country's government, it is quite possible that the quota rent accrues to the private sector of the exporting country or to any other than the importing country. For it may well be that international traders in fact dominate trade to such an extent that they are able to take advantage of the gap between the higher price the importers are prepared to pay, and the lower price the exporters are willing to accept. So without institutional knowledge about the international trading system, it is very hard to find out who really captures the quota rent.

Indeed, as far as the timber market is concerned, there are signs that the (inter)national logging companies and timber traders exert a significant power on the market, so that they can succeed in capturing the larger part of the difference between market prices and production costs (Repetto, 1989; Binswanger, 1991). This power often is fairly concentrated, and one is for instance reminded of the fact that over 95 per cent of the value of African timber is exported to the EU.

Moreover, governments of the producing countries commonly are in urgent need of foreign exchange, and will therefore be reluctant to exert too much pressure when they try to increase their share of the quota rent. All this leads to a fairly strong competition at the supply side of the international timber (products) market and to a so-called buyers' market.

- Governments will seldom be prepared to transfer the tariff revenues – or for that matter quota rents – fully to the other country's government or private agents. After all, if the importing country's government receives import levy revenues, that it should subsequently transfer fully back to the exporting country's government (or vice versa), why should it not leave the task of collecting the tariff revenues to the exporting country itself?
- With respect to the timber trade the issue of where any timber tariff or surcharge should be levied – at the importer's or the exporter's side – has become an issue. If tariff revenues are used to finance a fund for promoting sustainable timber production, the question boils down to the issue of who manages the fund and how it will be managed and controlled. If funds are managed mainly by the importing countries (e.g. through FAO's TFAP or through ITTO) its operation might become fairly costly and – in conformity with TFAP experience – research rather than sustainable management oriented. Moreover the distribution of the fund between the various producers might become a matter of dispute. If, instead, the responsibility to manage the fund is left mainly with the exporting countries, monitoring the allocation of the funds might become an issue (specifically if the funds are used to subsidize sustainably producing timber concession holders), as well as the burden of the administrative overload on the national administrative capacity.
- Even if one restricts the analysis to a rather straightforward product, such as tropical timber, the market quickly becomes fairly complicated and segmented. Consequently, the tariff structure of levies on timber and on timber products (or for that matter the quota structure) is not only fairly complex itself, but leaves room also for setting up a tariff structure in such a manner that disincentives are created to move timber processing capacity to the producing countries. The technique used is to apply a cascading import tariff structure, whereby the import tariff size increases with the domestic value added component. In the EU, USA and Japan, for instance, there are no, or only small, import levies on roundwood; levies are somewhat higher for processed wood products, and still higher for finished timber products. This is

clearly meant to discourage processing activities in the tropical regions, and explains why the importing countries generally resist the idea of imposing uniform (i.e. charged on both unprocessed and processed timber) import tariffs, or of introducing differentiated surcharges aimed at invalidating the cascading tariff structure.

- The behaviour of both suppliers and consumers may be affected by the application of trade policy, and in different ways, depending on the trade instrument applied. A quota, for instance, may affect the competition regime at the supply side, and therefore shift supply curves, etc. The impact may be quite different, if, instead, a tariff is imposed.

Supply, demand and substitution elasticities
If a protectionist measure is taken, irrespective of whether this is a tariff or a quota, its impact on volumes and prices will very much depend on the supply and demand elasticities. In the most simple case – represented by a partial analysis based on a model with linear supply and demand schedules – this can easily be illustrated with the help of some figures.

In Figure 2.3, case 1, supply is elastic (supply curve flat) and demand inelastic (demand curve steep); in case 2 the opposite holds. p* is the equilibrium price before the tariff is imposed; p is the price received by the exporters, and p+t is the price paid by the consumers. q* is the volume traded internationally before the tariff is imposed; q is traded thereafter. So q∗t is the government's tariff income.

The difference between the two cases is clearly illustrated by the figures. Whereas in case 1 the terms of trade loss of the exports is fairly small – timber prices on the international market are hardly affected by

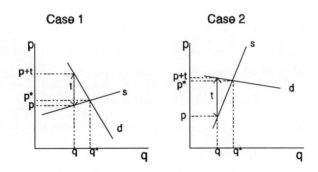

t = tariff
p = price
q = quantity
s = supply
d = demand

Figure 2.3 Supply and demand elasticities, and the terms of trade effect of a tariff

protectionism – the opposite holds in case 2, where the consumer hardly will notice that a tariff is levied, because the bill for the tariff revenues in fact is largely payed for by the exporting country. So, the elasticities of supply and demand will decide which of the countries will actually pay for the tariffs, irrespective of whether the tariffs are imposed by the importing or the exporting country (note what the country which does impose the tariff affects, and which country actually cashes in tariff revenues, however).

The question therefore arises: what can be said about timber trade elasticities? Based on a literature survey, NEI (1989) argues in this respect that the price elasticities of consumer demand for tropical timber probably lie anywhere between –0.5 and –1.2. It has been argued, however, that import demand for tropical timber is to a large extent only indirectly related to consumer demand, because timber is mainly imported by the timber processing industries. For that reason McKillop and Wibe (1987), for instance, argue that demand elasticities should be estimated through a combined factor and consumer demand model. Thus far it has remained unclear, however, how the choice for such an alternative model might affect the size of the demand elasticities.

In an ITTO study carried out by SBH in the Netherlands (1991, pp.67–70), a model was developed wherein import demand for tropical roundwood and sawnwood were treated separately. One of the main conclusions was that the demand for roundwood turned out to be fairly inelastic, whereas sawnwood demand, instead, was rather sensitive to price changes.

Another important factor, except of demand elasticities, is the elasticity of substitution, i.e. the degree to which alternatives for tropical timber will replace timber on the basis of relative prices. Earlier quantitative studies, carried out by McKillop and Wibe (McKillop, 1980; Wibe, 1984) showed that substitution elasticities are fairly small. More recent qualitative information based on inquiries of timber trade entrepreneurs, indicates, however, that the issue of potential substitution of timber producers by other products should not be underestimated. ITTO (1989), for instance, stated 'that although many tropical timbers maintain a strong share in market sectors which emphasizes the intrinsic characteristics of the raw material, few are uniquely free from substitution by alternative materials'. With respect to the assessment of substitution elasticities, the same applies as with respect to demand elasticities, namely that a distinction should be made between substitution at the industrial level on the one hand, and at the consumer level on the other hand.

Just as with (import) demand and substitution of demand elasticities, the picture of supply elasticities of timber is unclear. One major complicating factor is that elasticities are defined on the basis of a given *ex ante* behaviourial pattern, whereas these patterns tend to change themselves. So, with respect to supply elasticities the issue is to distinguish between shifts of the supply curve on the one hand and of shifts along the supply curve on the other. Some researchers have argued that there is no *a priori* reason why timber supply curves could not have a negative sign, and therefore bend backwards. The argument would be that governments more or less are in control of timber supply through their issuance of concessions, and that they can use their policies vis-à-vis the concession holders to acquire foreign exchange through them. A lower timber price would then act as an incentive for the governments in the tropical countries, at least in the longer term, to introduce a more lenient policy with respect to the issuance of timber concessions. The impact might be an increasing timber supply.

The impact of an import or export tariff on timber on the producing country would be disastrous, however, if the price elasticity of supply indeed were negative, specifically if supply is more inelastic than demand. This case is represented in Figure 2.4.

As one can easily deduce from Figure 2.4, a tariff t will lead to a new market equilibrium whereby the timber volume traded has expanded instead of contracted as a result of protectionism! Moreover timber prices have collapsed to such an extent that even the price including the tariff t is now smaller than the original equilibrium price, p*. In fact the suppliers through their substantial terms of trade loss, not only fully paid for the tariff revenues, but are forced to an even larger reduction of their timber prices. The producers therefore are in all respects worse off, even if tariff revenues would fully accrue to them.

One can easily infer on the basis of Figure 2.4 that if supply is totally inelastic or given (S-curve becomes vertical), the terms of trade loss for the exporters would exactly equal tariff revenues.

With respect to the final impact of any kind of surcharge on the market equilibrium, two more general aspects should be borne in mind:

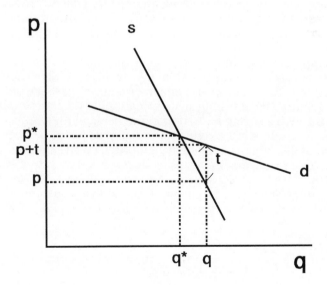

Figure 2.4 The impact of a tariff on timber trade in case of a negative price elasticity of supply

1) The cost of import timber is only a small fraction of the final mounted price, and so an increase in the price of tropical timber, say with an added surcharge, will therefore generally not strongly affect final demand. So, a small surcharge of 1–3 per cent will not lead to a serious decline in demand for tropical timber, because its impact will be lost in the mêlée of normal price changes caused by, for instance, exchange rate changes, market conditions and government regulations (NEI, 1989, p.67). There is some evidence that demand may fluctuate if trade measures are introduced based on speculative motives, however. The imports of tropical timber by the Netherlands, for instance, sharply increased in 1989, because of fears that the Malaysian government would introduce a serious increase of its export levies. These actions had a short-term impact on prices and stored volumes, but did not affect long-term demand, however.

2) The revenues resulting from a small surcharge would remain modest, even if the surcharge were to be levied on all OECD tropical timber imports. On the basis of 1988 OECD timber import values and assuming the surcharges would have no impact on trade volumes, the surcharge revenues would amount to some \$70m per percentage point surcharge per annum (some \$35m if applied to EU only).

Shift in the supply and demand schedules
Demand schedules can shift – as a result of preference drift, or due to altered trade, or because of changes in the consumers' value systems, or as a consequence of policy incentives, etc. One could, therefore, envisage a scenario wherein publicity and consumer actions increasingly contribute to a feeling among the general public in the OECD-countries that timber is a somewhat contaminated product if guarantees (e.g. through a hallmarked-system) that it has been produced sustainably are missing.

For example, in The Netherlands, some municipalities already forbid timber to be used in construction taking place under their direct responsibility; in addition a consumer action has started at the beginning of the 1990s aiming at stopping timber logging (*kappen met kappen*) through a strong reduction in the demand for tropical timber, as well as demanding more rights for indigenous people. Other private environmentalist organizations have emphasized the options of debt-for-sustainable management swaps or debt-for-nature swaps (Centrum Hout, 1988, p.10; UNESCO Sources, October 1990, p.14). Thus demand shrinks autonomously, and demand schedules consequently shift downward. At the export supply side, shifts can equally apply, for a number of reasons:

•   foreign exchange needs;
•   increased competition among the timber producers;
•   a more restrictive government policy;
•   new investments in logging activities;

- new timber species for production are discovered;
- an increasing domestic use, and so on.

It is fairly clear that where both supply and demand schedules themselves are in constant flux, a comprehensive analysis of the impact of trade policy devices is very hard to carry out. Basically one requires a simultaneous rather than a partial model to make a meaningful assessment. The sections on modelling in the following chapters will deal with such an approach.

3

TROPICAL FOREST MANAGEMENT

INTRODUCTION

Tropical forests can be managed for a number of reasons, such as the production of wood and other forest products and services (e.g. ecotourism), soil and water conservation or for the conservation of natural values. According to FAO (1993, p.47), forest management includes the demarcation and setting aside of a forest area for the production of forestry goods and services, an inventory of site and stand, the preparation of a management plan and control, and (eventual) application of silvicultural treatments. Forest management is, however, often dominated by timber production incentives for its cash generating capacity and, until recently, little attention has been given to managing the non-wood products and services of tropical forests. In this chapter, therefore, a survey of current management practices for the production of tropical timber in Africa, Latin America and Asia will be presented (see pp.49–55) The extent to which tropical forests are managed for the conservation of biological diversity, and natural and associated cultural values will be discussed on pp.56–58.

Most of the tropical forests are publicly owned. Governments in tropical timber producing countries therefore play two main roles in the management of their tropical forest resources: directly, by constituting the rules for protection or exploitation in any particular government-owned tropical forest; and indirectly by setting the legal framework and through policies which influence the forest management practices of other forest resource users. In many cases, however, government stewardship of tropical forest resources has not been very successful. The factors discouraging sustainable management of tropical forests will be discussed on pp.59–67.

Although the extraction of tropical timber for international trade plays only a relatively small role in the destruction of tropical forests (see Chapter 1), the direct and indirect effects on deforestation are not insignificant. Many of the present initiatives to combat deforestation in the humid tropics concentrate on the international trade of tropical timber. The efforts to reduce deforestation via international trade incentives, even if the extraction of tropical timber for international trade plays only a minor role in deforestation processes can, among others, be explained by the fact that timber extraction is the factor which can be most readily corrected and whose destruction is considered to be the least necessary (ESE, Vol.1, 1992).

The most important initiative in this respect is that in 1990, representatives of both producer and consumer countries, organized in the International Timber Trade Organization (ITTO), committed themselves to 'Target 2000'. The general acceptance by the ITTO members of Target 2000 means that from the year 2000 onwards only tropical timber derived from sustainably managed forests can enter international trade. The ITTO initiative will be discussed further on pp.67-68.

FOREST MANAGEMENT PRACTICES FOR TIMBER PRODUCTION

In order to gain insight into the current status of forest management in tropical timber producing countries, ITTO made detailed investigations, the results of which are laid down in several volumes by continent. A summary of these results can be found in Poore et al. (1989). According to the ITTO:

> the area of tropical moist forest which is demonstrably under sustained yield management in the member countries of the ITTO (excluding India) amounted, at the very most, to about one million ha and has now been reduced by about one fifth on environmental grounds. This is out of an estimated total area of some 828 million ha of productive forest remaining in 1985 in all the countries in which it occurs.
>
> (FAO, 1993, p.50)

It should be mentioned, however, that ITTO applies a rather strict definition of management for sustainable timber production: it should be practised on an operational rather than an experimental scale; it should include the essential tools of management defining objectives, working plans, felling cycles, yield control and prediction etc. and it should meet the wider political, social and economic criteria without which sustainability is probably uncertain. For an elaborate discussion of the elements of sustainable forest management and definitions the reader is referred to Poore et al., 1989, Chapter 1.

Tropical Africa

According to Rietbergen (in Poore et al., 1989, pp.40-73), almost all of the timber produced in Africa derives from natural forests which are being logged without a management plan or from forests which are, either planned or unplanned, converted to other uses (i.e. agriculture). Rietbergen conducted a study in six African tropical timber producing countries (Cameroon, Congo, Ivory Coast, Gabon, Ghana and Liberia) and found that in the West African countries studied most of the tropical moist forests have been logged over at least once and that in the other countries only remote and inaccessible forest areas are still intact. Since timber exploitation in this remote area is not economically feasible in the short term, it can be concluded that the current system of unplanned forest exploitation will have to give way to timber extraction from managed natural forest, supplemented by timber derived from other sources such as managed secondary regrowth, agroforestry and plantations, in the near future.

Although long considered as an alternative, plantations cannot fully replace natural forests as a source for tropical timber since a) the establishment and maintenance costs of plantations are high and investments in plantations are of a high-risk nature as compared to those required for the management of natural forest, and b) because plantations can not yield the same natural values or the same variety of natural products as natural forests and the quality of the timber produced is often inferior to the timber derived from natural forests. Since maintenance could no longer be financed, plantations in several African countries have been abandoned.

The current practice in Africa can be described as 'mining' the forests for timber, where little or none of the benefits obtained from timber production are reinvested in the forests in order to ensure future yields. Rietbergen discusses a number of reasons for this destructive exploitation method such as forest and land-use policy and its application, the economic conditions under which timber exploitation takes place, and the lack of available information to forest management.

Although in more accessible areas close to suitable seaports African tropical forests are more intensively exploited than those in more remote areas, forest exploitation is generally very selective, with removal rates of 5 to 35 metres per hectare whereas in Southeast Africa these rates vary from 50 to 120 metres per hectare. The reasons behind the low African exploitation rates are that the African tropical moist forests are less uniform and contain less commercial species, access and removal are more difficult, and the capacity of domestic markets to absorb the commercially less valuable species is in general small. In more accessible areas, however, due to the disappearance of prime species by overcutting, export bans on those species often combined with tax incentives and rising prices, and forests are increasingly logged for lesser-known species.

The very selective harvesting system in Africa has some important disadvantages. Since most of the commercial species need a lot of light to grow, for regeneration such species need gaps in the canopy which would be provided by a more heavy and intensive logging system. Furthermore, since for a given volume of timber a comparatively large area has to be logged over, relatively large areas of forest are opened up for shifting cultivators for a relatively small volume of timber. Because logged-over forests in general remain unprotected, the selective harvest system has led subsequently to the deforestation of large areas in Cameroon, Ivory Coast, Ghana, Liberia and Congo.

Governments of African countries have increasingly become aware of the need for more sustainable management of their natural forests and in several countries measures have recently been taken to improve forest management. Most countries thereby stimulate an extensive form of management of large concessions by highly capitalized logging and processing enterprises. Although most countries have committed themselves to a more sustainable form of exploitation of their forests, they fail to apply existing policies and legislation. In many cases agriculture is perceived to be more important than forestry and agricultural policies are developed without considering the impact on forestry. The bias towards agricultural interests also leads to the tendency to treat forests as a convertible rather than a renewable resource. As a consequence, existing forest protection rules are not enforced, the funds required for forest protection are lacking, and offenders against forest legislation are not, or not sufficiently, penalized.

The control of logging activities in Africa's tropical moist forests is also insufficient. Loggers are not being equipped with the necessary instructions to limit felling damage (climber cutting, directional felling, marking of residuals) in order to ensure the advance growth of commercial species after harvesting. Moreover, insofar as logging rules and regulations exist they are often not applied since forestry departments fail to enforce them.

Although policies to improve forest utilization are commonly introduced, these policies often remain without the intended effect. Incentives to increase the use of lesser-known species, for example, have only had occasional success. In most countries, however, the major share (50–80 per cent) of the volume harvested still comprises only one to three of the well-known species. Policies intended to create employment by stimulating domestic processing generally led to unnecessary wastage of timber due to inefficient production methods and a loss of tax revenues. Tax incentives to improve logging practices generally remain ineffective since taxes are either too low or of a too general nature. In cases where national logging enterprises are favoured, already inefficient forest utilization is further increased, since these enterprises are not being equipped with the necessary credit and training facilities to improve the efficiency of their operations.

The improvement of forest management in Africa is furthermore constrained by the general economic circumstances (indebtedness, inflation, overvalued currencies, etc.) and the situation on tropical timber markets (price fluctuations, uncertainties). General development problems and the uncertain timber market situation leave little room for governments to make extra efforts to improve the management of their forests. More information is required for the development of effective land use, forest, economic and financial policies, but this information is generally lacking since, for the same reasons as above, research does not have a high priority.

In summary, forest management in Africa, despite current measures and a rising awareness that forests need to be managed more sustainably, is in general absent or failing. According to Rietbergen, sustained yield management in humid tropical Africa is only practical if a number of conditions (firm forest policy and legislation, adequate forestry and forest management research and training, efficient logging control, sufficient forest fees which are reinvested in the forestry sector, adequate land-use policies, etc.) are fulfilled simultaneously. Otherwise the tropical moist forests in Africa will continue to be mined.

Tropical Latin America

The present status of forest management among ITTO members in Latin America (Brazil, Bolivia, Ecuador, Honduras, Peru and Trinidad and Tobago) has been studied by Synott (Poore et al., pp.74-116). The situation in these countries differs from Africa in that, according to Synott, many elements of natural forest management have already recently been introduced such as the legal declaration of forest reserves, inventories, silvicultural measures, botanical and ecological research, harvesting licences, in concession terms and some management plans. All member countries succeeded in setting aside forest reserves for productive forestry, national parks and other conservation areas. In all of these countries the forestry department takes an active part in many forest management operations, often with foreign assistance.

Although the need for adequate legislation is recognized and all countries have an elaborate package of forestry legislation and policies currently under revision, these laws and policies are seldomly effectively applied. In all countries, existing forest legislation is rarely respected or enforced.

The logging system, as in Africa, is highly selective. Only a few species are taken and logging intensity averages 2-5 metres per ha. Stumpage fees vary widely (from 0.55 US$ per cubic metre in Brazil to $US 7.50-40 in Honduras (Poore et al., 1989, Table 4.9, p.110)) due to some extent to delays in revising laws and regulations, but also to existing differences in attitudes of national governments towards how their forest resources should be exploited. The issuing of logging permits or concessions also varies widely: where in Bolivia only 195 concessions cover 22 million ha of tropical forest, in Trinidad as many as 800 loggers have been issued with licences to exploit the 75,000 ha Tropical Forest Production reserve. Timber processing is, in all member countries, mostly carried out by small, technically inefficient but profitable sawmills.

Forest inventories have frequently been carried out but are suspected to be unreliable. Moreover, where inventory results are available, they often remain unused by forestry services or timber producers in the planning of concessions or logging operations. Most of the available forest research is on botanical composition and ecological processes. With respect to forest management research, some important silvicultural experiments, although little coordinated and related to management reality, are carried out in Brazil and Trinidad. In many areas, furthermore, plantation and enrichment trials are carried out and permanent sample plots for management purposes, research and treatment trials have been established in Trinidad, in several areas of Brazil and in Honduras. All countries (except Trinidad) have training facilities for forestry staff at university level, but although the number of graduates currently exceeds that required for current management and logging control practices, it is far below the number required to manage the Latin American tropical forests effectively. Training facilities for lower level forestry staff are in general inadequate or even absent.

As opposed to Africa, in Latin America some control by forestry services over logging operations is exercised with the issuing of licenses, in the terms of concession agreements and by occasional roadchecks of log transports. Despite the introduction of these measures forest management in the Latin American ITTO member countries can not be characterized as sustainable. According to Synott only in the 75,000 ha Production Forest Reserves of Trinidad are production operations managed for sustained yields. In other important cases (Tapajós National Forest, Brazil, von Humboldt National Forest, Peru) attempts to arrive at integrated forest management have not (yet) succeeded. Despite the absence of sustainable timber production operations, timber production possibilities in Latin America are still not seriously endangered. Vast areas of forest have up till now remained unlogged and will, even after logging or temporary clearance, continue to yield timber. In those areas where timber resources are declining, the awareness of concessionaires, politicians and the forestry service staff that forest management should be improved is increasing, which gives good prospects for more sustainable production practices in the near future.

In order for forest management to be fully sustainable the following measures should be included in forest management operations:

• advance planning of the location and intensity of the annual cut;
• control of the actual cut with regard to the planning;
• protection of the timber exploitation area to prevent encroachment and/or illegal logging.

In practice, even in areas where forest services seem to have complete control by means of concession licences, the payment of fees and the application of regeneration rules are lacking. In many areas even control on logging activities is absent. Only a few concessions, in cases where the threat of timber supply shortages is already felt, is forest management not far from sustainable. But this does not hold for most of the Amazon region, where reservation, protection and the control over logging operations are lacking.

Most of the forest land in the countries studied is publicly owned. In most countries the government also has legal control over trees on private land. The majority of the publicly owned forests are, however, not managed at all. Forestry services are only responsible for the management of the remaining part of the tropical forest area which has been reserved either for production or protection. Even in the reserves forestry services generally lack the means to take active control. In the other forest areas the possibilities and means for forestry services to exercise control are even less. Although all countries reserved substantial areas of the tropical moist forest as conservation areas and forest reserves, many of these areas have been encroached upon by agricultural settlers. As in Africa, in many countries agricultural interests often are given higher priority than forest management and protection, and agricultural agencies act independently of the forestry services. In many cases forest conversion and settlement is encouraged by the political elite.

The prospects for sustained management in Latin American production reserves will, according to Synott, depend primarily on the willingness and capability of national governments to control commercial logging effectively, to coordinate the activities of the various government agencies concerned with land use, forestry and forest conservation, and to give higher priority to the needs of those living in or near the forests.

Tropical Asia

The present state of forest management in ITTO member countries in Tropical Asia (Thailand, Malaysia, the Philippines, Papua New Guinea and Indonesia) has been described by Burgess (Poore et al., pp.117–209). Most of the forest resources of this region are located in Indonesia: 65 per cent of the total forest area; 80 per cent of the production forest; and 90 per cent of all virgin stands. In Indonesia, as well as in Malaysia, it is government policy to reserve permanent production and protection reserves. In Thailand and the Philippines, however, forests are regarded as residual state land, since these countries have no active forest reservation policies. In Papua New Guinea most of the forest lands (97 per cent) are privately owned by either individuals or clans and the opportunities for the government to constitute permanent reserves are restricted to limited areas only.

Although greatly varying in proportions of forest under concession, terms and period of concessions, and the selection of concessionaires, in all the countries studied, logging operations are covered by concession agreements. According to Burgess, almost all of these forests under concession agreements are, at least nominally, managed. Forest management in these areas is generally adequate with regard to prescribed limits to yield or coupe, completeness and orderliness of fellings and confinement to coupe boundaries, and the residual stand. In other respects (protection of residual stand, silvicultural work, the following of felling cycles with no relogging in between, the maintenance of logging roads and post-felling erosion measures, the protection of unworked forest, and the writing and enforcement of working plans) management remains, in general, insufficient. Of these inadequacies the protection of unworked forests and residual stands seems, according to Burgess, to be the most serious.

In Latin America and Africa, the forestry departments in the Asian countries are, in general, not seriously understaffed. In view of high forest revenues, governments (except in Papua New Guinea) have in recent years greatly increased their forestry department staff. Exploitation monitoring and control by the forestry department staff seem to be more efficient in smaller logging operations. Larger operations are usually controlled by the concessionaires themselves and the government in these cases assumes that these concessionaires have an interest in sustainable management and most large operators hinder the possibilities for the forest field staff to carry out a thorough investigation of their activities. As a consequence of the lack of support by the government, forestry field staff in many cases are highly dependent on the goodwill of a concessionaire, who often does not hesitate to take full advantage of this situation. An adequate control of logging operations by the forestry department staff is furthermore obstructed by politicians with interests in the timber industry and by the interference in forestry department matters by influential government officials of other departments.

Most forestry departments have, in the past decade, seen a substantial increase in their budgets. They therefore, with the possible exception of Papua New Guinea, seem adequately facilitated. The Asian region, as opposed to Africa and Latin America, is well endowed with (well-known and reputed) forest research institutes. In several places university courses for senior officers are available. In general, forestry department officers as well as the forestry field staff are adequately trained.

The most important factor which prevents sustained-yield management in the region is the illegal clearing of forests, in particular of worked-over coupes, by shifting cultivators and in some countries (Thailand, the Philippines) by illegal loggers. As a consequence of enabling individual politicians to issue licenses and increase cutting rates, for their own political or financial interests, the second most important factor which prevents sustainable management of the forests in the region is the opening up of forests which should remain closed and overintensive logging in concession areas.

Although, in general, techniques and information for sustainable forest management are available and forestry departments in the Asian region are adequately staffed and funded, the size and facilities of the field staff are in some cases lagging behind. In Indonesia, for instance, more than half of the professional foresters employed by the forestry department are stationed in Jakarta. Of further concern is the status of the field staff: a forest official should be able to meet the manager of the concession on which he is stationed on equal terms and should not, which often is the case, be dependent on the concessionaire for his housing, transport, etc.

The country with the most favourable conditions for sustainable forest management in the region is Malaysia. Malaysia has a strong forestry department and a highly trained and disciplined field staff, outstanding research institutes and is guarding its forests well. In Malaysia individual politicians do not have the power to issue licences or to influence cutting rates. Moreover, the establishment and status of the National Forestry Status makes forestry a matter of national importance which cannot easily be interfered with or overridden by officials of other departments.

FOREST MANAGEMENT FOR CONSERVATION

Protected or conservation areas

Most tropical countries recognize the need to protect natural habitats, but few agree on the extent of that protection. However, the number and extent of areas under 'International Protection Systems' have risen over the years. Unesco (1978, p.511) distinguishes between the following general values and objectives to explicitly protect ecosystems, e.g. by the creation of forest reserves:

- the preservation of large, relatively self-contained natural areas in each of the world's major ecosystem types to safeguard the evolutionary processes;
- the protection of representative areas to safeguard the great diversity (biological and geological) of the biosphere's natural ecosystems, and to assure the preservation of the genetic pool they contain;
- the assurance that the normal regulatory functions of the biosphere can continue without irreversible disruption;
- the provision of representative and unique natural systems where basic and applied environmental research and education can be carried out, where baseline and monitoring studies can be initiated, and from which ecologically-sound planning and management of land and water resources can be derived and applied;
- the protection of watersheds, particularly from erosion and downstream sedimentation: this will also help to maintain high quality of water and to prevent serious flooding;
- the protection of fish and wildlife;
- the protection and conservation of plant species and their values as recreational resources, timber products, sources of genetic material for plant breeding, medicinal plants, climatic enhancement and ecosystem regulation;
- the provision of a wide spectrum of undisturbed areas for aesthetic and recreation purposes, and the development of modest natural-area based local tourism economies.

Based on these arguments virtually all tropical countries with serious forest resources have in the meantime created protected areas, such as parks and related reserves, although only rarely has this been achieved in the context of overall rational land use planning. 'Conservation' or 'protection' areas can be defined as areas of land managed through legal or customary regimes so as to protect and maintain biological diversity and natural and associated cultural resources (FAO, 1993). Table 3.1 gives an overview according to WRI, 1990, pp.292–293 and pp.300–301.[1]

---

[1] Although the FAO (FAO, 1993, pp.48–49) carried out a special study on the management of forests for conservation purposes, the results of this study are only available on regional level and could therefore not be used in this study.

Table 3.1 Forest resources and protected areas in the humid tropics in the 1980s
(in 1000 ha unless indicated otherwise)

	Forest area	*Area protected**	*(% of national land area)*
Latin America	785,875	56,015	
Bolivia	66,760	4837	4.5
Brazil	514,480	20,096	2.4
Colombia	51,700	5614	5.4
Ecuador	14,730	10,619	38.4
Guyana	18,695	12	0.1
Peru	70,640	5483	4.3
Surinam	15,000	735	4.6
Venezuela	33,870	8619	9.8
Africa	463,928	43,622	
Angola	53,600	890	0.7
Benin	3867	844	7.6
Cameroon	23,300	1702	3.6
Central Africa	35,890	3904	6.3
Congo	21,340	1353	4
Côte d'Ivoire	9834	1958	6.2
Eq. Guinea	1295	0	0.0
Gabon	20,575	1753	6.8
Ghana	8693	1175	5.1
Guinea	10,650	13	0.1
Guinea-Bisseau	2105	0	0.0
Kenya	2360	3095	5.4
Liberia	2040	131	1.4
Madagascar	13,200	1031	1.8
Nigeria	14,750	960	1.1
Senegal	11,045	2177	11.3
Sierra Leone	2055	101	1.4
Tanzania	42,040	11,913	8.5
Togo	1684	463	6.7
Uganda	6015	1332	13.4
Zaire	177,590	8827	3.9
Asia	272,826	21,409	
Fiji	811	5	0.3
Indonesia	116,895	14,067	7.8
Kampuchea	12,648	0	0
Lao Dem. Rep.	13,625	0	0
Malaysia	20,996	1101	3.4
Myanmar	31,941	173	0.3
Papua Nw. Guinea	38,175	7	0
Philippines	9510	521	1.7
Solomon Islands	2440	0	0
Thailand	15,675	4677	9.1
Vietnam	10,110	858	2.6

Source: WRI, 1990, Table 19.1, pp.292–293, Table 20.1, pp.300–301.

* National protection systems usually include sites listed under international protection systems, and combine natural areas in five World Conservation Union management categories (areas are at least 1000 hectares; access is at least partially restricted): scientific reserves and strict nature reserves; national parks and provincial parks; natural monuments and natural landmarks; managed nature reserves and wildlife sanctuaries; and protected landscapes and seascapes.

Table 3.1 clearly illustrates that the total area protected in most of the 40 countries considered varies between (almost) zero and some ten per cent (with the exception of Ecuador with a high share of 38 per cent). Countries that rank high (>7.6 per cent, but <15 per cent) are, for instance, Tanzania, Senegal, Venezuela, Thailand, Togo, Indonesia and Benin; hardly any protected areas (<0.3 per cent) can be found in Equatorial Guinea, Guinea, Guinea-Buissau, Kampuchea, Lao Democratic Republic, Papua New Guinea, Guyana and Myanmar.

Although the above data suggest that a significant proportion of the tropical forests is under management for conservation or protection, it should be borne in mind that legal protection alone can hardly ever ensure actual protection. Limited resources, lack of political will and other factors often limit the amount of enforcement possible. Because of all this the effective protection of rainforest reserves is an extremely difficult task. Many people, officials as well as private individuals, have great financial interest in forest exploitation.

> Logging companies are often willing to pay corrupt officials for a concession to extract trees, or to pay guards not to control or patrol a given area within a nature reserve. High timber prices for certain species may result in the illegal extraction of trees from a reserve or even a national park. In this way, the original value of a reserve can be so adversely affected that it is eventually abandoned.
>
> (Jacobs, 1988, p.243)

A comparable sad story can be told with respect to the feasibility of removing illegal subsistence farmers from the forest reserves, or even from preventing the mass of landless people from encroaching on the area.

Moreover, even when an area is effectively protected, activities just outside its boundaries can have a severe impact on its fragile ecosystems. One should, in this respect, also recognize that it is vital that the size and inaccessibility of forest reserves be such that the abundant diversification of natural species is preserved; small, cut-up forest complexes either do not guarantee this at all or only ensure it to a lesser extent.

The minimum size of a rainforest reserve

Much uncertainty exists as to what the minimum size of a rainforest reserve must be to survive in the long run. In determining the minimum size, the main factor according to Jacobs (1988, p.245) would seem to be the retention of animals in the forest ecosystem, while the plants must be present in a sufficient quantity to maintain heterogeneity and thus their adaptability to possible change or disease. The best available estimate he gives (see also Marshall in Whitmore, 1977, p.145) is that of some 1000 to 25,000 individuals, depending on the species, with about 5000 as a likely average. When this is related to a tree population density, true for many species, of less than one tree per hectare, often one tree per 3–5 hectare, one can calculate that to maintain the viability of an individual tree species an area of

150-250 km² is needed, providing that all the land is suitable and the tree population is evenly distributed, both of which are unlikely. Whatever the estimate, one has to think in terms of hundreds of square kilometres.

> For mobile animals like ungulates, monkeys, hornbills and other large birds, however, this is not a great distance to cover, and consequently such animals can stray out of the reserve and may or may not return. If they leave, their 'work', the dispersal of the larger seeds, comes to a standstill . . .. The vital process of exchange of genetic material will then break down, slowly, imperceptibly . . .. In fact, the reserve's area needs to exceed the range of the most mobile animal which plays a vital role in the ecosystem. If a reserve is too small to contain all its original inhabitants, various processes will set in, i.e. invasion of secondary species, on the one hand, and extinction on the other. The forest as it existed in its original form will be replaced by something quite different, less varied and therefore less useful for mankind at least in the long run.
>
> (Jacobs, 1988, pp.245-246)

FACTORS DISCOURAGING SUSTAINABLE MANAGEMENT

Introduction

None of the countries in the regions reviewed in the previous section practises sustainable forest management to a full extent. Before arriving at a discussion of the factors discouraging sustainable forest management a few remarks concerning the characteristics of tropical rainforests and their environment are in order.

The tropical rainforest, with its many species and rich interacting structure is dynamically fragile. Although the rainforests are well adapted to persist in the relatively predictable environment in which they have evolved, they are much less resistant to disturbances by man. It is very difficult to regenerate a tropical rainforest. This is due to a limited seed dispersal and highly specialized requirements for germination and seedling establishment. Methods that are being applied to regenerate tropical rainforests are generally derived from those used in temperate forests and are therefore usually insufficient for a tropical rainforest.

In Jacobs (1988, pp.4-6) the following problems related to the characteristics of the rainforest ecosystems are mentioned:

- In comparison to other forests, their potential for utilization is greatest in terms of quality, but smallest in terms of quantity.
- They cannot easily be cropped in quotas.
- They contain huge capital assets in the form of timber which, unlike non-timber products, generally cannot easily be harvested without serious ecological damage.
- They cannot be 'managed' without the loss of many species.
- They are unusually fragile and, once damaged, do not recover, or recover too slowly for any human planning.

- They are so complex that even their main ecological features are poorly known even to professional decision-makers and misunderstandings about them abound.

Sustainability is first of all difficult to define due to its multifaceted character. Therefore, in order to achieve sustainable production, a large number of constraints have to be overcome. In Table 3.2 a list of the potential factors influencing the government to support non-sustainable logging and farming in the tropical rainforest is presented (Dogsé, 1989).

Conditions and factors discouraging sustainable tropical forest management

At least 80 per cent of the forest lands in the humid tropics are state-owned. Royalties and other terms in timber concession agreements are therefore, to a large extent, the responsibility of the local governments. According to Repetto (1989 and 1990; see also pp.20–21) most governments in tropical timber producing countries capture only a

Table 3.2 Potential factors influencing government support for non-sustainable logging and farming in tropical rainforests

---

Vast forest areas
Deforestation = 'development'
Initially large supply of tropical timber
International demand for tropical timber
Debt crisis
Lack of capital for investment into sustainable technology
Possible export revenues from timber, crops or cattle
Easy way to capitalize forests by selling concessions for logging, export crops or cattle
Little or no government investments needed
No obvious management alternatives with similar short term incomes
Guided or forced by creditors or aid agencies
No assessments of forests' total value
Transfer of industrial country forestry and agricultural techniques not suitable in the tropics
Growing populations
Growing domestic demand for food and fuelwood
Transmigration of growing urban populations to reduce pressures in the cities
Poverty
Unemployment
Military security, difficulties of controlling uncleared tropical forest areas
Non-democratic political systems; e.g. groups affected neither politically nor economically powerful
Soft bureaucracies; opportunities for personal benefits to individual politicians
Government's lack of power of enforcement or control
Weak or non-existing government environmental departments
Lack of co-ordination between departments and activities (no holistic approach)
Weak environmental NGOs
High discount rate, favouring short-term benefits
Lacking knowledge of environmental response to human impacts
Time lag differences; immediate revenues, negative environmental impacts take time
Aggregated/synergistic effects; each impact may be marginal, sum non-marginal
Wish to master (hostile) nature

---

small fraction of the substantial value, or economic rent of the mature closed-canopy tropical forest of which they are proprietors. Governments typically lease timber lands not through competitive bidding, which would give them a larger share of the rents, but on the basis of standard terms or individually negotiated agreements. These governments obtain only a modest share of the exploitation value of their resources because royalties and fees charged to timber-concession holders are kept low, export taxes on processed timber are reduced to stimulate domestic industry, and income-tax holidays to logging companies are granted.

Precisely because these governments allow most of the resource rents to flow to timber concessionaires and speculators, who are often linked to foreign enterprises, both domestic and foreign entrepreneurs – many with little forestry experience – have been attracted to the search for quick fortunes. Governments, under pressure from timber companies, have granted timber concessions that cover areas far greater than they can effectively supervise or manage and that even in some cases extend into protected areas and national parks. The prospects for excessive profits have attracted politicians as well: in Thailand, Sarawak, Sabah and the Philippines, for example, cabinet ministers, senators and other senior politicians are involved in the timber industry.

One could argue that governments in the tropics furthermore fail to invest enough in stewardship and management of the forest. Although sustainable forest management is in many countries formally laid down in forestry codes and in concession agreements which ensure sustained productivity over at least several cycles, in practice almost nowhere are forests being managed sustainably.

The perverse incentives established for timber companies by the terms of many concession agreements, actually discourage loggers from managing their concession area sustainably. Even though intervals of 25 to 35 years are prescribed between successive harvests in selective cutting systems – and longer intervals in monocyclic systems (when all saleable timber is extracted at once) – most agreements run for 20 years or less, some even for less than five years. Levied fees are often simply based on the volume of wood extracted. This method encourages 'high-grading', a practice in which loggers take out only logs of the highest value and do so over large areas and at a minimum cost, in which case extensive damage is often inflicted on residual stands (ibid).

The reason why no greater efforts are undertaken to reform current revenue collection systems are given by Palmer (1994; in Poore et al., pp.164–167). According to Palmer past attempts to reform have often failed, are unpopular with politicians, logging companies, buyers and even the treasury itself, and advocates of past reform policies have been put under severe pressure. The failure of governments to manage their forests in a more sustainable manner is, according to Palmer, also a consequence of the status and organization of their forestry services. Most forestry services lack planning and development sections, and where planning and development activities are undertaken they tend to concentrate on short-term problems, rather than on long-term policy in order to avoid conflicts with logging companies and cattle ranchers who often have considerable political power. Forestry department officials are furthermore often insufficiently skilled to compete with the representatives of other departments for the limited available funds.

In his analysis of current management practices of tropical forests Palmer (1994; in Poore et al., pp.154–189), besides the above institutional factors also discusses a number of other factors (population growth, social and cultural characteristics, land tenure systems, technological development, economic valuation, forestry staff education, information gaps) which discourage sustainable forest management. Firstly the increase in the number of people living in the forests and their use of traditional land management methods results in considerable forest losses, whereas these methods are sustainable for a smaller number of people. Forest farmers in many cases do not place much effort in improving their methods to arrive at long-term settlement, since they either regard farming as a means to generate cash for another (urban) occupation, or they lack the information or financial means necessary to make such an effort worthwhile. Not only has the number of people making a living in the tropical forests increased but also the area which each family clears annually. This increase is a result of the availability and widespread use of the chainsaw, the general decline in the trade of minor forest products and the increased market for land. Current land tenure systems also often result in the loss of forests: clearance is often regarded as an 'improvement' to be rewarded with property rights which can be sold.

Technological developments have also altered production methods in the logging industry: the chainsaw, tractor and logging truck make it possible to harvest logs over a large area using only a small number of people (Palmer, 1994; in Poore et al., p.162). In most areas it is highly profitable to invest in these large-scale mechanized operations since payback periods on the materials are short with only a small risk involved and, as a consequence of the failing forestry policies of many national governments, raw materials are cheap and controls are non-existent or easy to avoid.

Another factor which leads to inadequate forest management and forest loss is the inability of tropical foresters to come up with valuation methods of forest goods and services which could be used as a strong argument for more sustainable management of a country's forest resources (ibid, pp.163–164). It is, however, difficult to arrive at such methods since renewal times are long and uncertain, and the value of indirect products such as soil conservation, a continuous flow of clean water and climate stability, is hard to estimate. In general, the role of forestry in countries with tropical forests remains small since forestry staff are insufficiently trained and educated in land-use management and development planning. Courses tend to be limited to the teaching of basic science and a more multidisciplinary approach is mostly absent, or not accompanied by the necessary field studies and practice.

Sustainable forest management is furthermore discouraged by the lack of information available for land-use and forest management planning (ibid, pp.169–170). Information is required on the productive capacity of land under different exploitation forms. Although efficient methods for the static inventory of forest have been developed, and actual inventories for the planning of concessions and logging operations are not uncommon, data from dynamic inventories are very scarce. The lack of forest productivity estimates leaves national forest services at a weak position in the land-use debate. According to Palmer (p.172) the loss of a considerable part of national permanent forest estates can be traced back to the inability of foresters to provide land-use planners with adequate data on productivity and multiple benefits of the tropical forests.

The industrialized countries have contributed to the above forest policy problems in the tropics. European and USA companies are involved in logging and processing activities, especially in tropical Africa and Latin America, but Japanese businesses now dominate in the tropical timber trade (see pp.35–37). As log supplies were successively depleted, Japanese firms have shifted their activities from The Philippines to Indonesia, then to Sabah and Sarawak, and now they are interested in Amazonian forests. Increased logging activities in Amazonia can be expected because Brazil is seeking funding from Japan to complete a road that would connect the Amazon with the seaport of Lima, Peru, on the Pacific Coast, opening a number of new markets for Amazon timber and other products (WRI, 1990, p.105). According to Repetto (1990) the Japanese firms generally do not manage their holdings in a sustainable manner; the common practice is to harvest as much and as fast as possible. Moreover, Japanese firms have participated in the bribery, smuggling and tax evasion going on in tropical timber exporting countries.

Many countries encourage processing rather than the export of logs by export restriction measures on logs to increase both employment and the value added to forest products. They must provide strong incentives to local mills to overcome high rates of protection against the importation of processed wood in Japan and Europe. Measures such as bans on log exports or export quotas have created inefficient local industries. The protection measures can create powerful local industries able to resist regulation.

The above situation is changing, however, because many countries are actually taking steps to capture resource rents at full value. Furthermore, a number of governments are strengthening their forest-management capability with the help of development-assistance agencies. The World Bank and the Asian Development Bank are planning to provide loans for forest-management improvement for a dozen countries. Most of these loans will be used to support forest-policy reform as well as institutional strengthening. In the framework of the Tropical Forestry Action Plan more than 50 countries are preparing national action plans to conserve and manage their forests.

An example of a country trying to increase the forest area under sustainable management is Peru.

> In Peru the forest is harvested in long narrow strips designed to mimic the gaps created when a tree falls from natural causes. The strips to be harvested are carefully selected, and animal traction is used to avoid soil compaction. Harvesting can thus be done without serious environmental damage, and the regeneration that takes place is rapid, abundant, and diverse. This experiment is being conducted in collaboration with the Yanesha Forestry Cooperative, a group of indigenous people who own the land communally.
>
> (World Bank, 1992, p.149)

With the current unsustainable production methods, resources such as forests, arable land and clean water are becoming scarce and must therefore be managed more effectively if the benefits derived from such resources are to be sustained. Forests under the current conditions are only partly renewable, and even more serious, it is possible that on some lands sustainable production methods can only be reached at very small, and therefore unprofitable extraction rates. This reinforces the need to use the resource as efficiently as possible and to achieve maximum revenue generation from those resources used. Among the areas where efficiency improvements are required are the following:

- **Log extraction.** Logging as currently practised damages significant amounts of standing timber, due in part to the fact that the concessionaire is less interested in future values from the timber left than in immediate gain.
- **Timber recovery.** After initial selective logging, very little timber is recovered from areas which are cleared for agricultural development. If the timber in the conversion forest is not recovered, the total national output will be significantly less than it might otherwise be.
- **Timber processing.** Sawmilling in Indonesia recovers about 43 per cent of the log compared to 55 per cent in comparable developing countries. Thus, improvements in sawmilling operations alone would increase saw-timber exports and associated revenue generation by 28 per cent from the same volume.
- **Plantation development.** Appropriately sited plantations could produce ten times more low-grade timber than an equal area under natural forest, thus providing timber for domestic construction and preserving more valuable tropical hardwoods for more specialized functions. But investment in plantations is not now attractive to concessionaires who have access to an almost cost-free resource in the natural forest.

The economic and ecological sustainability of logging the rainforest through successive timber extraction cycles has, for example, been demonstrated in the Malayan Uniform System and its modifications. However in other cases the situation is less clear. Problems include the degree of damage done during each logging operation (particularly to the soil), the slow rate of recovery growth of commercially valuable species after logging, and the hidden subsidies to the timber extraction industry. It is often suggested that most rainforest logging is a one-off activity with no foreseeable second or third harvest. Fears are also expressed for loss of habitat and species due to forest disturbance; loss of potential for non-timber products such as cane, fruit and chemicals; and the impact of humans using fire and cultivation on forests after logging is completed.

Logging activities are generally carried out by private firms under terms specified by the government, which may stipulate the duration of the concession rights, the logging practices to be used, and the obligation of the logger to provide for postharvest treatment of the forest. Although these terms have been stipulated, forest agencies have been unwilling or unable to enforce them. Due, among other things, to a lack of transport and good maps, forest agencies are unable to discover the value of the resource they protect. Examples are available in many countries where forestry officials who attempted to reinforce restrictions have been assaulted and even killed (World Bank, 1992). These risks do not contribute to the enforcement of the restrictions on logging activities.

A problem with concession agreements is the short period of the concession. Most of the concessions are of a relatively short term, compared with the regeneration period of the tropical rainforest. Therefore, incentives have to be built in to ensure regeneration. Longer-term contracts or contracts with provisions for performance-based extensions can force concessionaires to bear the costs that their initial harvests impose on future resource returns. They also permit concessionaires to reap the future rewards that are the necessary incentives for good harvest and regeneration practices. Furthermore, it is important that the concessionaires are ensured that their property will be respected by the local people. It is therefore necessary that these local people have the incentives to protect the forest. As long as the benefits of timber production do not (partly) accrue to these local inhabitants, their only way left to take (illegal) advantage of the land is to use it for shifting cultivation purposes.

The major problem of implementing any sustainable management system is each system's difficulty in combining its current competitiveness with its future sustainability. In the literature rough estimates are made about the costs of substituting the current supply of unsustainably produced timber for its sustainably produced equivalent. In NEI (1989), the World Resource Institute states that US$1 billion is needed for at least five years for the conservation of the tropical forests and another US$1 billion per year for a period of fifty years for the preservation of biological diversity. The International Bank for Reconstruction and Development states that US$5 billion is required for the sustainable management and conservation of the Amazon region alone (NEI, 1989).

During the last few years the international institutions have been trying to change the economic conditions for sustainable management.

The World Bank, for instance, now includes forestry components in rural development, in agriculture, in energy, in industry, and in area development projects. In its 1991 policy paper on forestry, the World Bank reflects the growing urgency of the deforestation problem, and it urges both a better understanding of the causes and implications of degradation, and a recognition that bank lending for other projects (particularly agricultural settlements and infrastructure) has sometimes had an undesirable impact on forest resources.

In tropical moist forests, the World Bank will encourage governments to adopt a precautionary policy toward the use of forest resources, since there is limited scientific knowledge about these ecosystems and biodiversity and since further work is required to establish viable sustainable management systems for commercial logging. Specifically, the bank (World Bank, 1990) will not finance commercial logging in primary tropical moist forests. Also:

> financing of infrastructure projects . . . that may lead to the loss of these and other primary forests will be subject to rigorous environmental and social assessments. In all countries, and for all types of forests, Bank lending operations in the forestry sector will distinguish between projects that are clearly environmentally protective (such as reforestation to protect watersheds), or socially oriented (e.g. farm and social forestry), and all other forestry operations (e.g. commercial plantations). The first two types will be assessed in terms of their social, economic, and environmental merits. The third type will require government commitment to conservation and sustainable use of forest resources, as well as to an improved policy and institutional environment
>
> (World Bank, 1990)

In addition, in order to manage the forests more sustainably, the governments should raise taxes and royalties to achieve more rational rates of extraction and improve public revenues. There are, however, practical problems in capturing existing rents where incentives are low for accurate reporting and monitoring. In principle, a number of different approaches could be taken (World Bank, 1990):

- **Competitive bidding.** The governments might require competitive bidding for the next round of long-term concession rights.
- **Stumpage tax.** A greater proportion of revenues could be generated as stumpage tax, a tax on the total volume of standing timber whether or not harvested, thus reducing the tendency to distort or under-report harvested volume.
- **Export taxes.** Taxes could be collected primarily on export products. This does not provide a mechanism for full rent collection from timber going into domestic consumption and, unmodified, could distort the pattern of production for domestic versus international markets.

According to Sharma and Rowe (1992), a multisectoral approach is needed in making the transition to sustainable use of forest resources possible. Development efforts and operations in other productive sectors, such as agriculture, energy, and industry, should promote forest conservation by minimizing destructive deforestation. An important step will be for governments to change the existing incentives that encourage uncontrolled destruction of forests. This will involve the elimination of current distorting policies that caused the incentives for excessive logging, and the correction of market failures by defining property rights more clearly. Furthermore, the governments should create private incentives for planting trees, for practising preservation, and for establishing sustainable management systems for timber production. At the same time, governments will need to devise proper land use policies (including land tenure arrangements) that clearly spell out how forests are to be used. For example, different areas of an intact forest could be set aside for:

1) conservation of biodiversity and environmental services,
2) production of timber and non-wood products,
3) conversion to sustainable agriculture,
4) restoration, and
5) multiple uses.

Research is needed on sustainable management of the natural forest, possibly including improved silviculture techniques and enrichment planting. Further work is also needed to increase the volume and quality of plantation timber. Additional attention needs to be given to production and marketing of secondary forest products which will encourage smallholders to utilize, not convert, forested land: and to mixed agroforestry systems which will help restore the diversity of resources available for human use. Further attention should also be given to the environmental implications of logging.

Finally, however, the most fundamental task to be fulfilled, if progress towards sustainable management is to be made, is that at a global scale, the externalities are internalized, and are thus reflected in the timber prices.

THE ITTO INITIATIVE

Although only a small part of the annual loss of forest resources in tropical countries as shown in Tables 1.1a–c can be attributed directly to timber extraction for international trade, its share is significant, especially if the indirect effects of opening up the forests for other users is taken into account. Moreover, according to ESE (1992, vol.I, p.1), forest destruction caused by logging companies operating on international markets is 'the least necessary and can be most readily corrected'. The rapid destruction of tropical forests in the past decades led to concern among tropical timber–producing and timber–consuming nations that if no measures are undertaken the supply of merchantable tropical timber will decline rapidly and ultimately become unavailable. For many tropical timber producing countries this will mean a loss of an important source of foreign currency earnings and government revenues. In 1990, the tropical timber exporting country members of the ITTO committed themselves to Target 2000: from the year 2000 onwards tropical timber exports from these countries will be limited to timber originating from sustainably managed forests. In order to arrive at this goal these countries furthermore committed themselves to the implementation of the ITTO guidelines for sustainable management and to provide annual progress reports which will allow for the monitoring of a country's achievements towards meeting Target 2000.

Whether all internationally traded timber in the year 2000 will actually be derived from sustainably managed forests in the full sense depends on the one hand on whether ITTO members manage to agree on the interpretation of the ITTO sustainable development objective and on the other hand on the capacity and willingness of producer countries to implement all necessary measures. ITTO defines sustainable forest management as 'the process of managing permanent forest land to achieve one or more clearly specified objectives of management with regard to the production of a continuous flow of desired forest products and services without undue reduction of its inherent values and future productivity and without undue undesirable effects on the physical and social environment.' (ESE, 1992, vol.I, p.1).

This definition contains three elements of sustainable development: environment, economy and society. Despite past efforts ITTO has not yet succeeded in formulating sustainable forest management practices in which economic development and environmental conservation are balanced to the satisfaction of all members. The degree to which the Target 2000 agreement should be regarded as a commodity agreement, an environmental or a development instrument still has to be resolved. Many producing countries are, moreover, strongly against the inclusion of well defined social objectives (Bass et al., 1992, p.46). It is therefore doubtful whether sustainable development in the full sense can be achieved by the agreement.

Whether Target 2000 can actually be attained is widely doubted. Current world trade in tropical timber amounts to about 70 million m^3 roundwood equivalent which, according to an ESE estimate (1992, p.2), would mean that in the year 2000, 70 million ha of forest will have to be brought under sustainable management. Considering that at present only 1 to 10 million hectares are sustainably managed and that, in general, forestry departments are ill equipped, understaffed and underfunded, Target 2000 in most producing countries seems almost impossible to achieve. According to Bass et al. (1992, p.47):

Target 2000 is both an inspiration and an embarrassment. It has stimulated a great deal of valuable activity, but is - if viewed strictly and objectively - almost impossible to attain, except perhaps if international trade in tropical timber were to cease altogether. It is also only truly valuable as a lever if it leads to the sustainable management of all production forests . . ..
A better measure of success would be the rate at which various nations move towards sustainable management.

4

FORESTRY AND TIMBER TRADE MODELS

INTRODUCTION

In the preceding chapters an overview has been presented of tropical forest resources, of factors underlying deforestation, the tropical forest products trade, tropical forest management practices and of existing policies and policy options determining or directed at deforestation rates or forest management practices. Before arriving at an integrated global simulation model in Chapter 7, in this and the following chapters partial models in which (elements of) the processes described have been formalized will be discussed. In this chapter, therefore, a review is given of existing forestry, forest conservation and timber production and trade models.

On pp.69-71 a literature survey on forestry and forest conservation models is presented. This survey, which was carried out during September 1989-mid 1990, concluded that only TROPFORM was able to simulate deforestation on a global scale. Several other models are available but all of them have their limitations; most of them are directed towards a particular forest area in one tropical country. Therefore, TROPFORM has been used as a starting point in the process towards modelling deforestation on a global scale. Pages 71–82 cover the economic relations underlying TROPFORM. Furthermore, some remarks will be made on the advantages and limitations of TROPFORM.

FORESTRY AND FOREST CONSERVATION MODELS: A LITERATURE SURVEY

In a literature survey Hoogeveen (1989) evaluated some forestry and forest conservation models on their usefulness for a global tropical forest simulation project. Since the main goal of the project is to develop a global simulation model in order to evaluate policy options, both from an international and a national perspective, and their impact on the tropical forest area, attention was therefore mainly focused on economic models.

In the literature, several examples of forestry and forest conservation models can be found. These models can be classified in two main categories:

1)   ecological models
2)   economic models.

Ecological models

In an ecological model attention is mainly focused on the forest as an ecosystem. This category of models has not been reviewed since the approach taken is beyond the scope of this study. One typical example may serve as an illustration, however. In Braat and Van Lierop (1987), three different ecological simulation models are discussed: the Nutrient Flux Density Model, FORCYTE, and the Spruce Budworm Model.

The main objective of the Nutrient Flux Density Model is to describe the main aspects of the dynamics of nitrogen in order to gain insight into the interaction between growth of biomass and nitrogen conversion. The model has been used to evaluate the impact of acid rain on Swedish forests.

FORCYTE simulates the long-term effects of a growing biomass and evaluates the possible policy options to counterattack the growing biomass. The Spruce Budworm Model has been developed in order to evaluate several policy options as an alternative to using insecticides in the Canadian forests.

Economic models

Economic forestry and forest conservation models are typically developed from the point of view of the economics of natural resources. The economics of natural resources are involved with the optimal exploration of natural resources (Howe, 1987). A distinction is often made between renewable resources, such as forests, and non-renewable resources such as fossil fuels.

Economic models dealing with the forest could be classified as forestry and forest conservation models. Forestry models are focused on the optimal exploitation of a forest from the point of view of the individual forest manager. Forest conservation models, however, are developed in order to conserve one or several forest areas, and are usually developed from the point of view of a regional or even higher government. Of the various models that have been developed in this field, most belong to the category of forestry models.

Forestry models

A wide variety of forestry models have been constructed from a theoretical point of view, of which the major part focuses on the optimal rotation period under different constraints. Examples of the latter type of models are found in Johansson and Löfgren (1985), Clarke and Reed (1989, pp.569-95), Crabbé and Van Long (1989, pp.54-65) and Kula (1988).

Several other models have been designed for practical purposes, e.g. in order to support the forest management in its decisions about the optimal exploitation of a specific forest. In this category three types of these models can be distinguished: wood production models, wood product models and wood processing models. Forest production models have been developed to (economically) optimize wood production in terms of costs, revenues and/or allocation. Examples can be found in Bare (1984, pp.1-18), and Braat and van Lierop (1987). In wood product models the main objective is to minimize costs or waste when converting wood into wood products (see also Bratt). Wood processing models instead are developed for analysis, control and/or optimization of the activities of people and machines when using wood for the production of specific products.

Forest conservation models
The major part of the forest conservation models – economic models developed for the conservation of one or more forest areas – are specified at the national or sub-national level. At the time when this survey was carried out (1990), only TROPFORM, a model developed by Grainger (1987), was specified at the global level.

Allen (1985, pp.59-84) describes a forest conservation model of the Dodoma region in Tanzania. The model, a multi-goal programming model, is used for comparing policy scenarios aimed at the transition from deforestation of public forest areas to the use of plantation wood. The model generates an optimal combination of efficient wood production, allocation of labour and conservation of the forest.

Hassan and Hertzler (1988, pp.163-168) describe a dynamic programming model for Sudan in order to develop an optimal policy to control forest exploitation as a source of energy on the one hand and desertification on the other hand. This model uses real prices of fuelwood, which take into account the costs of logging for future generations.

A linear deterministic simulation model of Mwandosya and Luhanga (1985, pp.1023-1028) models the impact of several policy options, e.g. growth of population and reduction in biomass demand per capita, on the deforested area in Tanzania. Gane (1986, pp.41-9) presents a sector simulation model, TIMPLAN, in order to determine the best strategy for forest development on a national level. The model generates, amongst others, projections of future demand for and supply of wood, costs and benefit in the forest sector, earnings in foreign currency and benefits for the national economy.

TROPFORM, the only model that can be used for policy analysis on a global scale, models several factors determining land use and leading (eventually) to deforestation. This model will be discussed in the next section.

TROPFORM

The future role of tropical rainforests in the world economy will diminish as a consequence of the current patterns of timber extraction (logging) and deforestation for agriculture. TROPFORM provides an economic framework to address these problems on a global level. It is a model to simulate possible trends in tropical rainforests; resources and

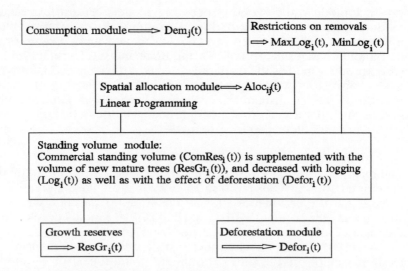

Figure 4.1 Flowchart of the TROPFORM model

tropical hardwood supply, consumption and trade. The model structure is shown in Figure 4.1 (see also Blom, Hoogeveen, and Van der Linden, 1990).

The central part of the model consists of a Linear Programming Model, the spatial allocation module in which the logging from the reserves of wood products from the different producing regions (21 in total) are distributed over the consumption of the world (eight macroregions). The reasoning behind this is that all demand for wood products is met given certain constraints which assure general inertia in the tradeflows, progress in the logging industry and a minimum amount of reserves. These constraints are of course influenced by government policies. Input in the spatial allocation module is the demand determined in the consumption module and the restrictions on removals. The removals, as an outcome of the spatial allocation module, together with deforestation due to the increase in farmland (deforestation module), the regeneration of previously logged reserves, the discovery of new reserves, and the maturing of intensive

Table 4.1 Timber producing countries and consuming regions

Asia-Pacific	Africa	Latin America	Macro-regions
Myanmar (Burma)	Cameroon	Bolivia	North America
Indonesia	Central Africa Rep.	Brazil	Japan
Malaysia	Congo	Colombia	Europe
Papua New Guinea	Gabon	Ecuador	USSR
Philippines	Zaire	French Guyana	Africa
Rest of Tropical Asia	Rest of Tropical Africa	Guyana	Australia &
		Peru	New Zealand
		Surinam	Asia-Pacific
		Venezuela	Latin America

plantations (growth reserves), determine the change in the reserves (reserves module).

Timber demand

In TROPFORM, demand from all major economic regions in the world and supply from the tropical regions are brought together; it explicitly deals with the economic processes which result in deforestation and in logging.

The model will here be discussed on the basis of the flowchart. (For the meaning of the symbols see also Table 4.2. The index j refers to the eight macroregions of demand; i refers to the 21 producing regions; t is time index (years), 1987 = 100.)

The central module of the model is the standing volume module. In its central storage equation the commercial standing volume of region i is kept up with the in- and outflows. The change in this stock variable is supplemented with the volume of new mature trees, and decreased with the level of logging, as well as with the effect of deforestation. The volume of commercially interesting trees is determined in the growth

Table 4.2 List of variables

Variable	Description
$Aloc_{ij}(t)$	Allocation of removals from region i to consumption region j
$ComRes_i(t)$	Commercial standing volume (m)
$Const_j$	Appropriate constant
$Defor_i(t)$	Annual area of deforestation
$Dem_j(t)$	Total demand in region j of the woodproduct
$DemPC_j(t)$	Demand per capita in region j
$ExpC_{ij}(t)$	Costs of international transportation
$ExtrC_i(t)$	Costs of extraction
Farmland	Farmland needed to support $Pop_j(t)$
$FoodPC_i(t)$	Food consumption per capita from own resources
$Forest_i(t)$	Area of forest
$GrFood_i$	Growth rate of food consumption per capita
$GrIncPC_i(t)$	Growth rate of income per capita
$GrPop_i$	Growth rate of the population
$GrYieldF_i$	Growth rate of yield per hectare farmland
$IncPC_i(t)$	Real income per capita
IncElas	Income elasticity of demand
$LimForest_i$	Minimum protected area forest
$Log_i(t)$	Wood extracted by logging
$MaxLog_i(t)$	Restriction on the removals $log_i(t)$
$MinLog_i(t)$	Restriction on the removals $log_i(t)$
$P_j(t)$	Real price
$Pop_j(t)$	Population of region j
$ResGr_i(t)$	Increase in reserves
$ResGrReg_i(t)$	Regeneration from previously selective logged-over forest
$ResGrNew_i(t)$	Discovery of 'new' reserves
$ResGrPlan_i(t)$	Maturing of intensive plantations
t	Time
$TransC_i(t)$	Costs of transportation to port
$YieldF_i(t)$	Average yield per hectare of all farmlands
β	Price elasticity of demand
τ	Measure of product substitution and consumer preference over time

reserves module. Three factors contribute to this increment: the regeneration of logged-over forest areas, the discovery of new standing volume and the maturing of intensive plantations.

The level of logging is constrained to a lower and upper level. These boundaries are determined by the restrictions on the volume of tropical wood logged, and are assumed to result from a general inertia in the market positions, a limited progress in the logging industry, and a finite amount of standing volume.

The actual level of logging is the aggregated solution of the LP module in the spatial allocation module, which results from the allocation of the product (in log wood equivalent) from the 21 tropical regions over the consumption of the eight macroregions of the world. Therefore this module gives the analysis its global dimension. The restrictions in the LP module are on removals, but also the demand has to be met. The level of demand is reduced if the level of commercial standing volume is insufficient to sustain the previous levels of logging for a longer period. Demand then becomes restricted by the conditions on the supply side.

In the consumption module wood also is expressed in log wood equivalent. The decrease in the commercial standing volume due to deforestation is dependent on the area deforested as a result of agricultural encroachment into the forest area (deforestation module).

Consumption module:

$$DemPcj(t)=Const_j*exp(\gamma*t)*IncPc_j(t)^{IncElas_j}*P_j(t)^\beta \qquad (4.1)$$

$$Pop_j(t)=Pop_j(0)*exp(GrPop_j*t) \qquad (4.2)$$

$$IncPc_j(t)=IncPc_j(t-1)*exp(GrIncPc_j(t-1)) \qquad (4.3)$$

$$GrIncPc_j(t)=GrIncPc_j(0)*exp(0.06*(incPc_j(0)-IncPc_j(t)) \qquad (4.4)$$

$$Dem_j(t)=Dem_j(t-1)*exp(GrPop_j+GrIncPc_j(t)*IncElas_j) \qquad (4.5)$$

In eq. 4.1. the most influential factors from demand theory for the individual consumption of a product have been selected:

1) the substitution of tropical hardwoods by wood or non-wood products or change in consumer preference over time (τ);
2) the income per capita ($IncPC_j(t)$);
3) the price level of the product ($P_j(t)$).

To simplify the analysis, two assumptions are made: first, no substitution of tropical hardwoods by wood or non-wood products or change in consumer preference over time (See also Grainger, 1986, p.185; Wibe, 1984). Secondly, prices are omitted from the consumption module, in common with many long-term forecasting models (Grainger, 1986, p.183). In fact, the regional average export prices did not show an upward trend over the period 1965–80.[1] Some

---

[1] The time series were obtained from Forest Product prices published by FAO and deflated with the index of export values of manufactured goods published in the UN

authors foresee such an upward trend, however (Repetto, 1990, p.20). What results is an individual demand function with a direct relationship between per capita income and per capita consumption of wood products.

To derive the total demand for a region, the demand per capita has to be multiplied by the level of population in year t (eq. 4.2). A simple exponential growth function is considered for the growth in population (Vu, 1984). For the income per capita also simple exponential growth is considered (eq. 4.3), but this rate depends on the general level of the per capita income compared with the per capita income in the base year (1987). This assumption (eq. 4.4) results in a smooth decrease in the growth rate in per capita income, because of the exponential functional form. The speed in the decrease is determined by a parameter set at 0.06, based on IIASA FSP/GTM.[2] From the above the total demand for final products by the end user can be derived according to (eq. 4.5) ((eq. 4.1) with $\tau = 0$ and $\beta = 0$, (eq. 4.3) substituted in this equation, and multiplied with (eq. 4.2)).

On the basis of 1981-1988 data the simulation output of the consumption module was compared with real-world observations in the case of Indonesia.[3] The 1981 data served as initialization of the consumption module: a growth rate of GDP per capita over 1980–81 (0.051), a GDP per capita for 1981 (RP 475,068 – 1983 constant market prices) and domestic sawnwood and plywood demand in 1981 (8,325,425 m³ logwood equivalent[4]), and an income elasticity (derived from Grainger, 1986, p.184). The sample mean for the reported reality over 1982–88 was 9,396,554 m³; for the simulations a sample mean of 9,376,662 m³ could be calculated, a difference of only 0.2 per cent. The sample standard deviation for the simulation output 512,882 was also close to the sample standard deviation of the reported reality 576,524 (11 per cent difference); it should be noted that the data referred to a time series. The correlation between the reported reality and the simulation output was high: R^2 is 0.94. These results warrant a certain confidence in the simulation output from the consumption module.

Monthly Bulletin of Statistics (UN, 1986), according to FAO custom. Regional averages were obtained by using hardwood log export values for the period from the 1980 FAO Yearbook of Forest Products. The world average of the log prices rose in 1979-80, but both remained fairly constant over the rest of the period. African prices fluctuated more with a sharp rise for 1972-74.

[2] The IIASA FSP/GTM model is a trade model of approximately 20 wood products requiring large amounts of data.

[3] In the period 1981-88 it was not expected that a reduction in demand due to the mechanisms on the restrictions on removals in the model would be necessary for the simulation output, because in 1987 the ratio between the commercial standing volume and the domestic demand was 259 (compared with the time horizon of the mechanism of 20).

[4] For sawnwood an average equivalent value for log of 1.7 and for plywood of 2.1 was used, calculated from Secretary General of Ministry of Forestry, p.138, 1990, Table II.10.

Restrictions on removals

The restrictions on the removals determine upper ($MaxLog_i(t)$) and lower ($MinLog_i(t)$) levels for the amount of logging per producing region, and represent a general inertia in the market positions of the regions, to assure limited progress in the logging industry and a minimum amount of standing volume.

The lower level is taken to ensure that domestic consumption in tropical regions is not crowded out by external demand. From this mechanism it results that the opportunities for domestic demand and exports are determined: these upper and lower levels cannot guarantee complete trade flow inertia. The incremental growth rate that is applied to the upper level assures progress in the logging industry: the capacity of the logging industry can not grow unlimited. The rate that is applied to the lower level assures that domestic demand can also profit from the greater opportunities in the domestic logging industry.

If the commercial standing volume has been exhausted to such an extent that current extraction rates are not tenable over the assumed horizons then the upper and lower levels will be reduced, each by its own mechanism. Due to this reduction the progress in the forestry sector can turn into a decline. As long as other suppliers can take over the external demand that cannot be satisfied by the producing region that has to cut down its production, this results in a shift in the trade flows, contradicting the general inertia.

If all the producing regions together cannot fulfil the demand from all macroregions, $MaxLog_i(t)$ is increased for the regions that have more than 15 years constant production ahead. If this is still not enough, the demand for the non-producing macroregions is decreased in the same manner as for the macroregions that do contain producing regions in case the $MinLog_i(t)$ is not sufficient for the domestic demand.

Spatial Allocation module:

$$MinZ(t) = \sum_{j=1}^{8} \sum_{i=1}^{21} (ExtrC_i + TransC_i + ExpC_{ij}) * Aloc_{ij}(t)$$

Subject to:

$$Log_i(t) \leq MaxLog_i(t) \quad \forall:i = 1..21 \tag{4.6}$$

$$Log_i(t) \geq MinLog_i(t) \quad \forall:i = 1..21 \tag{4.7}$$

$$\sum_{i=1}^{21} Aloc_{ij}(t) \geq Dem_j(t) \quad \forall:j = 1..8 \tag{4.8}$$

$$Aloc_{ij}(t) \geq 0 \quad \forall: = 1..21, j = 1..8 \tag{4.9}$$

$$Aloc_{ij}(t) \geq 0 \quad \forall:i=1..21, j=1..8 \tag{4.10}$$

The allocation of the supply of tropical logs from the tropical regions over the demand in the macroregions is modelled by an LP model. The decision variable is the allocation of the tropical wood products from the producing regions to the macroregions ($Aloc_{ij}(t)$). The sum of these flows over the destination determines the actual level of removals ($Log_i(t)$) in a producing region (eq. 4.6). The restrictions (eq. 4.7) and (eq. 4.8) result from the restrictions on the reserves module. (eq. 4.9) is the explicit assumption that all demand is met; (eq. 4.10) that all flows, the supply for the domestic or foreign market, are not negative. The costs considered in the objective function are specified as follows:

1) costs of extraction were considered to be a piecewise linear function of the density of reserves per hectare in the various regions;
2) costs of transportation to port were considered a linear function of the distance from the mean centre of country to the major coastal port;
3) costs of international transportation were estimated by considering a constant cost per m³ and per kilometre.

Deforestation module:

$$FarmLand_i(t) = \frac{FoodPc_i(t) * Pop_i(t)}{YieldF_i(t)} \tag{4.11}$$

$$FarmLand_i(t) = FarmLand_i(t-1)*exp(GrFood_i + GrPop_i - GrYieldF_i)$$

$$Forest_i(t) = Forest_i(t-1)-(FarmLand_i(t)-FarmLand_i(t-1))*(1-\\ exp(LimForest_i(t-1)-Forest_i(t-1))) \tag{4.12}$$

$$LimForest_i(t) = 0.1*Pop_i(t) \tag{4.13}$$

$$DeFor_i(t) = Forest_i(t)-Forest_i(t-1) \tag{4.14}$$

In this and the next two modules the analysis is confined to the 21 tropical regions. The deforestation module considers three types of land use:

1) farmland or agricultural land ($Farmland_i(t)$);
2) forest ($Forest_i(t)$);
3) the residual (only implicitly defined in the model) and which consists mainly of urban areas and non-forest nature, like swamps.

Considering a stable third category, then deforestation is defined as the encroachment of agricultural land. In the model the demand for food is the product of the food consumption per capita ($FoodPC_i(t)$) and the size of the population ($Pop_i(t)$). The total demand for food for a tropical region divided by the average yield per hectare in agriculture ($YieldF_i(t)$) determines the total amount of farmland that has to be used to satisfy the basic needs for food. Here the possibility of importing agricultural products is excluded; nor has a distinction been made between shifting and permanent agriculture. For the food consumption per capita, the size of the population, and the yield per hectare in agriculture, growth paths are considered with exogenous growth rates. If these results are substituted in the second part of eq. 4.11, the third part results. Agricultural encroachment, the increase in farmland, is determined as $Farmland_i(t)-Farmland_i(t-1)$.

The area of forest land is assumed to approach smoothly a limit which depends on the size of the population (eq. 4.12). To obtain this limitation the increase in agricultural land is restricted by a function in exponential form that is dependent on the difference between the lower bound to the forest area and the forest area itself. The speed of the smooth reduction is determined by this difference. In reality sometimes the percentage national forest cover fell to limits of below 10 per cent as population density (per ha) increased (eq. 4.13) (Grainger, 1986, p.116). The model implicitly assumes that the agroecosystem is stable after this limit has been passed. Deforestation is logically the reduction in forest land (eq. 4.14).

The land use categorization of the model makes no provision for land characteristics such as the variety of the slope of the area. There can be reasons to do so, however,[5] because logging on areas can cause erosion; similar processes of land degradation also hold true for clearing these areas for agricultural purposes.

In an Indonesian forest and land use inventory per province, excluding Java and Bali, Sutter distinguishes three slope classes: 0–8 per cent, 9–40 per cent and 41+ per cent (Sutter, 1989, pp.306–330, Table A17). The natural forest area in the 41+ per cent class can be used as a proxy of the area that in the case of deforestation would in any case suffer erosion. The percentage forest in this class to the total area per province remains smaller than 10 per cent in only three provinces: Riau, Sumatra Selatan and Lampung; between 11 and 25 per cent in seven other provinces: Sumatra Utara, Jambi, Nusa Tenggara Timur, Timor Timur, Kalimantan Barat, Kalimantan Tengah and Kalimantan Selatan; the share in the 11 other provinces is above 26 per cent. Therefore when the 10 per cent lower limit in forest cover of the model would actually be reached in these provinces they would almost all suffer from serious erosion problems. For Indonesia as a whole the percentage of forests on steep slopes is a formidable 34 per cent, corresponding with 60m ha (the ratio forest/population = 0.1 implicates 17.2m ha). However, 49m ha of forest in Indonesia is protected forest or parks and reserve forests, where no logging and agricultural encroachment is allowed, and 30.5 m ha of the permanent production forest is limited for production. The 60 m ha in the 41+ per cent slope class is 53 per cent of the total permanent forest. From the above it can be concluded that the limit in forest area considered in the model corresponds with a highly threatening environmental scenario.

---

[5] Armitage and Kuswanda for instance excluded the 46+ per cent category (slope > 45 degrees); see FAO/GOI in their projections of logging for Indonesia (1989, p.250).

Growth of reserves volume:

$$ResGr_i(t) = ResGrReg_i(t) + ResGrNew_i(t) + ResGrPlan_i(t) \qquad (4.15)$$

$$ResGrReg_i(t) = \frac{Log_i(t-40)}{4} \qquad (4.16)$$

where:

$ResGrNew_i(t)$ is a scenario variable;
$ResGrPlan_i(t)$ is exogenous; external projections are used in the model.

The standing volume of commercially interesting tropical trees can increase by:

- regeneration in logged-over forest areas ($ResGrReg_i(t)$);
- the discovery of new reserves;
- as a result of an expansion of the species considered commercial or of greater accuracy in estimating areas and standing volumes of forest ($ResGrNew_i(t)$); and
- the maturing of intensive plantations ($ResGrPlan_i(t)$).

Because the last two components are respectively a scenario variable and an exogenous variable, these components will not be discussed here.

In this module the three main assumptions are: a 40-year rotation period; 50 per cent of these logged-over forest areas have escaped agricultural clearance during the last 40 years; of the remaining forests a per hectare yield of 50 per cent of the original yield is considered.

The constellation of this regrowth mechanism together with the 15-year and 20-year horizons with respect to the restrictions on removals have also been inspired for each producing region by the silviculture systems they made obligatory to the logging companies in their forests at the beginning of the 1980s.

Commercial reserves module:

$$ComRes_i(t) = - \frac{ComRes_i(t-1) + ResGr_i(t) - Log_i(t)}{Forest_i(t-1)} * Defor_i(t) \qquad (4.17)$$

The remaining commercial reserves are determined by the bookkeeping of (eq. 4.17). In (eq. 4.17) the standing volume destroyed by deforestation is proportional to total forest area deforested. If deforestation has taken place, future logging in these areas is unlikely (except in a case where the deforested areas are turned into timber plantations).

Application possibilities: TROPFORM

After having specified the model structure of TROPFORM, the question arises which are its application possibilities. With the TROPFORM model three main trends can be simulated:

- simulate trends in consumption for tropical hardwoods;
- simulate the way in which this consumption is satisfied by different producing nations;
- simulate the effect of logging and deforestation on reserves and future supplies of tropical hardwoods.

The results of these simulations are dependent on the values which are given to the exogenous variables. However, in reality some of these exogenous variables can, at least partly, be influenced by governmental action. Other exogenous variables are fixed arbitrarily, and can easily be set to other values. This means that deviations from the base case, in which past trends are extrapolated, by changing some exogenous variables, can appear.

TROPFORM: Advantages and Disadvantages

TROPFORM was the only multiregional model to simulate future trends in the world's tropical rainforest reserves, and tropical hardwood supply, consumption and trade. In this section some advantages and disadvantages of TROPFORM as related to the required model for the TROPENBOS programme will be discussed.

For the TROPENBOS programme a dynamic multiregional model is required which gives at least the future deforestation rate for the different tropical regions and with which it is possible to simulate various policy options. Interdependency between different tropical and non-tropical global regions has to exist, among others, through international trade in tropical hardwood. Furthermore the model should have the capability to simulate the impact of wood and non-wood substitutes of tropical hardwood. In general the model has to represent those physical and socio-economic factors which are crucial for tropical rainforest reserves.

TROPFORM can, by means of scenarios, be used to test various policy options such as different supply and/or consumption conditions. With respect to model results, the model, as described in the reserves module, simulates the effects of logging and deforestation on tropical hardwood reserves and furthermore the contribution of intensive plantation, of regenerated previously logged forests, and of newly discovered reserves to future tropical hardwood supplies. In the consumption module, together with the spatial allocation module, trends in consumption of tropical hardwoods by importing and producing nations are described as is the way in which this consumption is satisfied by different producing nations and consequent long-term changes in major supply sources.

Important relationships between socio-economic variables and tropical hardwood consumption and reserves are described in the consumption and the deforestation module. The linkages which are required for the TROPENBOS programme are provided for in the following way. Future trends in consumption of tropical hardwoods are dependent upon income and population growth rates. The rate of deforestation is determined by a change in land use, which is in its turn determined by the need of an increase in food production as a consequence of population growth.

With respect to the number of relations and variables, TROPFORM is relatively small as compared with other global models and can without too much difficulty be converted to a PC version. For initialization and simulation purposes a limited dataset is required which is, especially in the light of the availability and quality of data on developing regions, an unmistakable advantage of TROPFORM.

With respect to the choice of modelling techniques a question arises. It is not quite clear why in the spatial allocation module a LP model is chosen to describe future international trade flows in tropical hardwoods. According to Grainger (1986) simulation does not seem the appropriate technique to model tropical hardwoods trade on a large scale. The advantage of optimization over simulation is that, although 'it is a less flexible technique than simulation, and demands more data, it can give more precise solutions. While simulation can indicate a "stable" solution, optimization offers a minimum cost or equivalent solution more compatible with economic theory' (Grainger, 1986, p.61).

Considering the model objective, to simulate, among others, possible trends in tropical hardwood trade, it is not clear why an optimimum solution is 'more precise' than a simulation result. The model objective is to provide a realistic description of future tropical hardwood trade patterns. An optimization procedure can very well generate trade flows in which the cheapest supplier of tropical hardwoods will be the only supplier until his reserves are exhausted; then a major shift will occur to a second supplier etc. This would hardly be a realistic outcome, given existing trade inertia as a consequence of all kinds of trade barriers. This is also acknowledged by Grainger, and such an extreme solution is prevented by means of maximum and minimum removals (MaxLog (t) and MinLog (t)) in the reserves module. However, as a consequence of the introduction of these restrictions the advantage of optimization of a 'precise solution' which is 'more compatible with economic theory' is lost.

TROPFORM does not contain a production module. Therefore relationships between tropical hardwood reserves and the production of final goods and services are not modelled. In a general basic model of natural resource use (Howe, 1987) the production function could be modelled as follows:

$$GNP(t) = f(L(t),K(t),R(t),t)$$

where:

GNP(t) = The production of final goods and services in region i in
           year t
L(t)   = Labour inputs
K(t)   = Capital goods inputs
R(t)   = Natural resource commodities
t   = Relative change over time as technology or other factors
           change.

With the help of such a production function, the effects of substitution between the different production factors, of a logging prohibition, or of technological change, could be simulated. In the model, possible future substitution of tropical hardwoods by wood or non-wood products is not explicitly visualized. In the consumption module a parameter (τ) in the individual consumption function gives the combined product substitution effect and consumer preference over time. In the total demand function this parameter is, however, assumed to be zero. Grainger (1986, Chapter 6) states that based on past trends and experiences the effect of substitution on the tropical hardwood trade is likely to be fairly neutral.

A final remark concerns a matter of consistency with respect to the representation of population growth. In the consumption module population growth is given by an exogenous growth rate (GrPop1) and in the deforestation module population growth is determined by means of a continuous logistic equation (dPop2/dt).

PART II

LAND USE AND SOCIO-ECONOMIC DEVELOPMENTS IN TROPICAL REGIONS

5

AGRICULTURE IN THE HUMID TROPICS

AGRICULTURAL SYSTEMS: SHIFTING CULTIVATION

The tropics provide a unique physical environment for agricultural activities. The key characteristic, except for a high potential for photosynthesis, the poor structure of many tropical soils, and the many biotic difficulties such as pests and diseases of many kinds, is the enormous variability: climates, soils and altitude differ widely and markedly, causing a wide variety of physical conditions even over short distances.

The consequent generally high costs of soil fertility maintenance – in tropical forest areas the main storehouse of nutrients is not the soil but the standing vegetation – explains the structure of agricultural production to a large extent, in particular the important role of shifting agriculture. Indeed, shifting cultivation is the most widespread farming system in the humid tropics (in terms of the number of people practising it), especially in the rainfed farming areas of low population density such as New Guinea, Borneo, Myanmar, Zaire, and the Amazon Basin. 'We have no basic idea of how numerous the shifted cultivator is: we simply do not know how many forestland farmers there are, beyond estimates that range from 300 million to 500 million. If the latter estimate is correct (it may even be an underestimate), he accounts for almost one in ten of humankind.' (Myers, 1991, p.24.)

A clear division of jobs between men and women is usually found wherever forest and bush vegetation must be cleared. In fire-farming in the forest, clearing is the man's job, while planting, weeding – if done at all – and harvesting are carried out by the women. The cut, burn, plant method of agriculture is the most common one reported. Typically, vegetation is cut usually towards the end of the dry season, allowed to dry for a while, and is then burnt as the rainy season approaches. The shifting agriculture system is equally popular in other tropical areas such as in the humid and semi-humid areas, or savanna and grassland areas in Africa and Latin America (see Figure 5.1).

On the following pages we will focus on the first category, i.e. shifting agriculture in the humid tropics, as much as possible. Since shifting cultivators are held responsible for over half of total deforestation, insights into the socio-economic and ecological sustainability aspects of this farming system seem imperative for virtually any attempt to contribute to reducing deforestation. Ecological sustainability of shifting cultivation systems implies the stability of the ecosystem involved, i.e. that agricultural activity through shifting cultivation is compatible with the tropical forests' carrying capacity. The restoration of soil fertility hereby plays a crucial role; with respect to shifting cultivation this is directly related to the length of the fallow period. If this period is long enough the soil will regain its fertility to a level that guarantees sustainability of the system; if the fallow period is too short, soil productivity will sooner or later decline.

Box 3 Definition and features of the shifting cultivation system

Shifting cultivation is the name used for agricultural systems that involve an alternation between cropping for a few years on selected and cleared plots and a lengthy period when the soil is rested.

Shifting cultivation is usually associated with the tropical areas. Its main features are:

- a rotation of plots;
- the alternation between short (mostly 1–3 years) occupation of plots and relatively long fallow periods, resulting in extensive landuse;
- for the cultivation of plots mostly fire is used, and furthermore almost exclusively human energy;
- cultivation of crops with a short cultivation period (one and two years);
- recovery of soil productivity – which has declined during the cultivation period – by means of spontaneous fallow vegetation;
- there are few pure subsistence holdings among the shifting cultivators; in most cases supplementary cash cropping is being practised.

A common method to classify farming systems is the characteristic R method. The characteristic R is defined as the proportion of the area under cultivation in relation to the total area available for arable farming. According to FAO one speaks of shifting cultivation if R is less than 33, i.e. less than 33 per cent of the arable and temporarily used land is cultivated annually. Otherwise one speaks of fallow systems (33<R<66) or permanent culture (R>66).

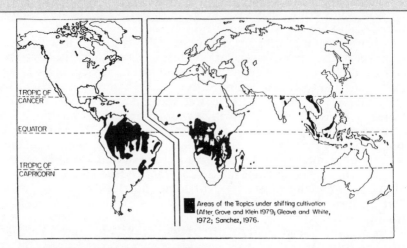

Source: FAO 1984, in Dorland et al. (1988).

Figure 5.1 Shifting cultivation areas in the tropics

Socio-economic aspects of shifting cultivation

Labour is by far the most important input in traditional shifting cultivation. It is provided by members of the family and its size is therefore restricted (especially during labour requirement peaks). The problem of seasonality further adds to the scarcity of labour. The most urgent need for high labour inputs often coincides with the period when fresh high quality food is most scarce, and low nutritional levels inhibit workers and draught animals from working to full capacity.

The role of capital is negligible in shifting cultivation. Although sometimes crop surpluses are bartered on a local market, virtually no capital is being invested in the cultivation system. In areas with low population densities and few other resources except for land, the farmers' choice for shifting cultivation can be regarded as rational because it is economically the best method of farming for them, requiring almost no cash inputs and very limited labour. All other farming systems require more inputs and more labour, and farmers will not turn to these higher cost methods unless it is essential and viable.

Naturally the crop yields of shifting systems vary greatly according to climate and soil type. For example, whereas on the acid soils of the Amazonas Basin 10 or more fallow years follow 1.5 years of cultivation, with the same rainfall on volcanic soils in Zaire there are only 2- or 3-year fallows for the same period of cultivation. In any case, however, the shortening of the fallow or the lengthening of the cultivation period beyond a certain point disturbs the equilibrium of the land use system. Shifting cultivation, which can be called 'balanced exploitation' where the fallow period is sufficiently long, becomes soil mining. Geertz (1963, p.26) argued that in the rainforests of Southeast Asia, in an ecologically-balanced shifting economy, no more than 20–50 people can live per square kilometre with guaranteed subsistence (Ruthenberg, 1980, p.61).

A second characteristic of shifting cultivation is related to the shifting cultivators' main target. Continued subsistence food production with a minimum of risk, instead of profit maximization, is often aimed at, for crop failure does not mean a financial loss but a threat to subsistence. To limit this risk, farmers sometimes may use more production factors including land than usually necessary for food production. Problems of uncertainty generally also reduce the flexibility of farm systems and their ability to evolve rapidly along apparently attractive development paths. These problems particularly apply where small farming units and low incomes make risk avoidance especially important. So, the problem of uncertainty cannot be separated from the low productivity of labour in tropical agriculture, whether measured in hours worked per available worker, or by physical performance per hour of work, or by the return per hour of work or per man-year.

Various cropping principles serve to reduce uncertainty: mixed cropping, which means the simultaneous growth of two or more intermingled useful plants on the same plot; phased planting;

intercropping (two or more useful crops in proximate but different rows); and crop rotations, i.e. alternating crop mixtures.

The above two factors, variability and uncertainty, to a large extent also explain the shifting cultivators' behaviour with respect to their role in deforestation. To give an example, burning, after the removal of forest vegetation, is a labour saving practice. The same effects as from burning could be obtained otherwise (e.g. by ploughing, the use of fertilizer, and weeding), but only with extra inputs of labour and/or capital. Another implication can be that the most fertile soils are by no means always preferred, precisely because this may require too many labour inputs per unit of production. Instead it can be more logical to turn towards the more marginal lands.

An example of the various labour inputs in rice production in some African shifting systems during the various stages of production derived from Ruthenberg (1980) is given in Table 5.1.

Table 5.1 Some examples with upland rice in African shifting systems (per hectare of first crop after clearance).

Country	*Sierra Leone*	*Liberia*	*Côte d'Ivoire*	*Cameroon*
Location	Bum	Gbanga	Man	Begang
Year	1971–72	1972	1974–75	1976–77
Technique	Traditional	Traditional	Traditional	Traditional
Method	Sample	Sample	Sample	Sample
Technical data				
Labour input (manhours)[a]				
Slashing and felling	135	254		115
Burning and clearing	79	164	301	184
Planting[b]	127	107	142	181
Weeding	145	37	292	416
Bird-scaring[c]	256	44	222	294
Harvesting	268	164	218	277
Threshing (4)	59[d]		84	224
Total	1069	770	1259	1691
Yield (t/paddy)	1.23	0.97	1.74	0.77
Economic analysis ($)				
Gross return (5)	112	122	203[e]	152
Material inputs				
Seeds	4	4	8	8
Tools	3	3	4	1
Income before interest				
and overheads	105	115	191	143
Income per hour of work	0.10	0.15	0.15	0.08

Source: Ruthenberg, 1980.

Notes: a) 6 hours per working day.
　　　 b) Including minor land preparation works. The land is not cultivated before planting.
　　　 c) Including hut building, fencing, etc.
　　　 d) The figure for Sierra Leone is an estimate from other surveys.
　　　 e) The figure for Cameroon is based on 1973–1974 prices.

Agricultural change

A key question arises, concerning what can be said about agricultural change: how do and can shifting cultivation systems evolve towards more classic agricultural production systems? Boserup (1965) clearly pointed out that agricultural developments depend strongly on factors such as population pressure, and classified the systems of land use with respect to degree of intensity - which is the ratio between the length of the cultivation period and the fallow period. Her classification ranges from *forest fallow cultivation* through bush and short fallow cultivation to annual and multicropping. The transition from one form to another is considered as a continuum, whereby the fallow is reduced to ultimately to negligible. The reasoning behind this was that a growing rural population will commonly have the effect of shortening the fallow period, e.g. short-fallow cultivators are likely to take annual cropping on part of their land. It is this transition that may call for the introduction of better ploughing, irrigation and weeding, in other words to adopt technological changes in the cropping system which result in higher yields per unit of land. However, two points can be made here. First, although examples abound of population-induced productivity growth through intensification, there are many other cases where intensification eventually has led to reduced labour productivity due to land overuse, and consequently soil depletion, exhaustion, even abandonment. But secondly and more importantly, intensification as a response to increased population pressure is not the only possible outcome. As long as expansion of the area under cultivation by (a) clearing more of one's own land or appropriating neighbouring lands or (b) migrating to other areas to develop new lands for agriculture, is an option, this can be the most rational and economic move for the shifting cultivator: 'The higher returns to labour from more extensive forms of shifting cultivation . . . is one of the main incentives for "forest encroachment" by shifting cultivators . . .' (Raintree and Warner, 1986, p.42).

In analysing the relationship between agricultural change and external factors, one realizes more and more that clear–cut answers to adjustment problems in shifting cultivation areas cannot be given nowadays.

> It is evident that no single theory can adequately explain the complex interactions between population growth and land use. It is thus illuminating to investigate the diverse patterns of adjustment in land relations which occur at particular stages of the demographic evolution. These adjustments will be more or less severe, depending on the nature of the tenure regime that determines access to land in a society, the level of technology, the availability of unoccupied land, and the public policy environment
>
> (Bilsborrow and Ogendo, 1992, p.43)

A more elaborate discussion of the economic factors which determine agricultural development paths will be presented on pp.90–100.

Various patterns of development

Various patterns of development from shifting systems towards permanent cropping can be distinguished in reality. One type of development, called *evolutionary* by Ruthenberg (1980; see also Geertz, 1963), evolves as follows: fallow systems evolve out of shifting systems if the planting of tree crops gives rise to stationary housing. More and more land is taken by permanent crops and food crops; less and less remains fallow. If the farmers start to live in permanent dwellings, or move only within a limited area, the shifting cultivation is generally supplemented by other major food-producing activities, including planting of tree crops, permanent gardens, permanent irrigation fields, and fish-ponds. Farmers now increasingly shy away from cultivating remote fields, because of the transport cost involved. Fields closer to the hut are preferred even if this means lower yields and more weeding. A further evolution of the system would mean permanent upland cultivation, with possible wet rice in the valley, based on an intensive fertilizer economy.

An example of *commercialized shifting cultivation* is that of manioc-flour producers at Castanaha, Para State, Brazil. Shifting cultivation in Para State occurs under the conditions of ample rainfall and ample land, but fragile soils. Rice, maize and manioc are planted after the clearance, the first two of which are minor products. The major objective is to obtain manioc which is processed on the farm into manioc flour, a marketable product. Eight to nine years of fallow follow 1–1.5 years of crops. Harvesting and processing occur continuously throughout the year. The farmer works with family and hired labour. Production is fully commercialized and more than a century of routine in the process has led to high labour productivities. Production occurs almost without purchased inputs. In terms of the efficiency in the use of support energies, this is one of the most productive systems of the tropics with an energy ratio of 65 units of food energy, ready for consumption as 'farinha' manioc flour, for each MJ energy input, in the form of human labour. Returns per hectare of land, however, remain low because of the lengthy fallow period (Ruthenberg, 1980, p.58).

Often traditional shifting cultivation can no longer be practised with, among other things, a fallow period long enough to restore soil fertility. The consequences are mostly far reaching, not only for the rural population facing increasing poverty, but also at a national level due to food shortages, increasing erosion, loss of natural resources, etc. In such circumstances technical solutions can only contribute to improvements if they fit within the socio-economic context and have political support to carry them out (Dorland et al. 1988, p.71).

So, with increasing population and growing production for sale, an adverse, involutionary instead of evolutionary pattern could also emerge. A number of distinct *stages of development* can be distinguished:

1) The oldest form is probably the shifting of cultivation and of the farming group in one direction within a primary vegetation.

2) According to the growing number of other claimants to land, the groups share the total area among themselves, then each family shifts in a circle in the area allotted, which is covered by secondary vegetation, and moves on after one or several decades. Relatively intensive cultivation is practised temporarily near the dwelling (R=30–50).[1]

3) With a further increase in population, instead of circular shifting with long-term intervals of fallow and cropping, the shifting of fields often occurs within a fairly limited area, with the tendency towards stationary housing.

4) A continuous increase in population and the development of cash cropping involve intensified shifting cultivation by shortening the fallow (R rises to 30–60). At this stage, the basis of land utilization has changed to such an extent that the system can no longer be regarded as shifting cultivation, but becomes a fallow system.

These processes usually lead to a situation of relative overpopulation on the land, which then results in lost soil fertility. Without changes in technology and inputs, only a small number of people can be supported compared with other land use systems.

The changing external conditions and shifting cultivation adjustments

The increasing incorporation of the various national economies in the tropical regions into the international economic system has led to an increasing integration of a shifting cultivation community into a broader society, economically, socially and politically. The almost closed economic systems of shifting cultivation of the past have opened up. This has contributed to an increasing need of shifting cultivators for cash income. Such income can either be generated from agricultural production for the market, or from off-farm employment. In the first case the area under cultivation can be increased and thereby the pressure on the fallow period. In the second case either off-farm activities take place during the off-season period, or they take place more permanently, often accompanied by migration to urban areas. This may further add to the common shortage of labour.

Degeneration of shifting cultivation systems of the involutionary type can eventually cause losses of natural resources. The increasing pressure on land leads to deforestation on a large scale, affecting the national economy for example through fuelwood shortages and erosion. Although the degeneration of shifting cultivation is not the only reason for deforestation, the search for a sustainable alternative for degenerated shifting cultivation systems could play an important role in preventing further deforestation.

A response of shifting cultivators to rapid population growth is usually a change in crops: crops of relatively high quality are replaced by lower quality crops that are less demanding in terms of land and/or

---

[1] See Box 3 for the definition of R.

labour. Also more manure and mulch is used. Fertilizer is either not available, not suitable under local circumstances, or too expensive. Especially with increasing population densities the need for manuring increases, but the possibilities decrease because the bush that supplies both grazing and mulch diminishes, and costs of transportation to more distant bush areas become prohibitively high. Furthermore, labour shortages might also restrict the possibilities for manuring. Finally, farmers can end up in a 'low level steady state' (Ruthenberg, 1980), wherein output is permanently low. Despite the intensification process – of land use as well as labour – the shifting cultivator goes downhill. Spontaneous intensification is not likely to occur as long as returns to labour are lower.

AGRICULTURAL CHANGE AND LAND PRODUCTIVITY

Agricultural change from the traditional shifting cultivation towards more 'advanced' systems of agricultural production, will usually involve a change in *land productivity*. How land productivity in a certain area will evolve is obviously to a large extent dependent on the agricultural system adopted and agricultural techniques applied.

Some broad categories of changes from shifting systems to other farming systems can be distinguished in practice:

- from shifting cultivation to permanent upland farming;
- from shifting cultivation to regulated ley farming, the replacement of the bush fallow by plants that are themselves useable, regenerate the soil productivity, and by allowing use of the plough;
- from shifting cultivation to perennial crops. The humid and semi-humid tropics are by nature forested, and the natural environment is well suited to perennial crops and to tree crops like bananas, oil-palm, rubber, cocoa, coffee, kola-nuts, and citrus;
- from shifting cultivation to irrigation farming, in particular to wet rice. Indeed, rice cultivation on the flood plains and valley bottoms has enabled Southeast Asian farmers to feed very large populations in most years;
- from shifting cultivation to artificial pastures;
- from shifting cultivation to ranching.

There is no commonly accepted theoretical framework that can explain how in a certain area the transition processes will evolve, and what alternative agricultural system will eventually be put in place of the traditional shifting cultivation system. However, an interesting theory on the impact of increasing population pressure on the agricultural system has recently been put forward by Bilsborrow and Okoth Ogendo (1992, pp.37–45). Contrary to earlier theories on the effects of population on agriculture and economic development, such as those of Malthus, Boserup and Davis, they basically argue that except for the 'classical outlets' of increasing population pressure on the land, an additional form of response should be distinguished: an increase of the land area under cultivation. Such an increase could not only be achieved by clearing more of one's own land or appropriating neighbours, but also by migrating to other areas to develop new lands for agriculture. Both of these forms of expansion of the land area under cultivation may be referred to as *land extensification*, and could involve deforestation.

They subsequently argue that there are four phases involved in the responses of land-use practices to population growth, each involving several adjustments, many of which are consecutive, others concurrent and some even cumulative. The four phases may be classified as *tenurial* (Phase I), *land appropriation* or *extensification* (Phase II), *technological* (Phase III) and *demographic* (Phase IV).

The proposition of Bilsborrow and Okoth Ogendo that the effects of population pressure on land use are likely to be felt first through changes in tenure arrangements (Phase I) is based on the facts that:

> (a) demand exerted on land by a growing population of necessity generates questions about individual and group entitlements to its growing scarcity, and (b) out-migration is in this stage usually not a desirable option. Being at the center of economic and social security in predominantly agrarian systems, the right of access to and control of land, i.e., tenure, is the key to decision-making about its use.

The second major phase of adjustment (Phase II) (which may occur simultaneously with the first or third stage) is land extensification, the appropriation of land rights, usually in frontiers not controlled by, or perceived as available to, a given community. This occurs as soon as communities no longer have a sufficient stock of land in their immediate area for present or future claimants through the adjustment mechanisms of Phase I. It does not matter whether the appropriation involves the acquisition of ownership or merely occupational rights through appropriation. Thus, in areas where there remains land which is both arable and physically accessible, extensification of agriculture may occur involving out-migration (rural-rural) of families to unoccupied lands and often land conflicts with existing, indigenous population groups.

The adoption of new technologies of land use (Phase III) is often a measure of social and cultural flexibility in society. It is not, therefore, always a response to external stimuli. For this reason, agrarian communities in Africa and Latin America have been known to accept and sustain radical changes in land use, e.g. the introduction of cash crops in place of subsistence crops, without fundamental changes in tenurial regimes. The distinctive feature of technological changes associated with population growth is, therefore, that it is accompanied by intensification, i.e. an increase in land productivity. Technological change represents, therefore, a more advanced phase in land-use management.

The demographic phase (Phase IV) is usually, in the absence of outside intervention, the last response in the population phase/land-use continuum. It involves fertility reduction through either (or both) postponement of marriage or reduction in marital fertility.

Obviously, the context directly influences what types of response occur. That is a major determinant of the extent to which (a) land tenure change, extensification or intensification of agriculture takes place; (b) the forms in which they occur; and (c) whether demographic factors play a significant role or not, including the following: natural resource endowments; institutional and attitudinal factors, deriving ultimately from the country's history (including prevailing land size and existing tenure arrangements and community controls); and government policies.

The above theory – and the statement in the last paragraph in particular – can be illustrated by the fact that the pattern of change in agricultural production depends inter alia on the degree to which one focuses on cash crops destined for export. Virtually no peasant society exists that does not sell a fraction of its production (Unesco, 1978, pp.427–431). As a result there is a considerable increase in land needs with its consequent pressure on the forests of the equatorial zone and on its margins. This is especially because some of the most important cash crop products are ones which only the tropical forest ecosystem can supply, such as cocoa, coffee, bananas and other fruit, palm–oil and natural rubber. These products are more varied and less easy to replace (by synthetic products or those of temperate agriculture) than the products of savanna ecosystems (mainly cotton and groundnuts). Generally they refer to woody perennial crops which can be grown within the natural ecosystem without radically changing it. In the most densely populated areas, whenever cash crops develop at the expense of food crops, a part of the money obtained is used to purchase food from outside sources. However, an important part of the expansion of cash crops orchards occurs in the areas of intact or little altered forest with a low or at least a moderate population. This adds to the severity of the deforestation problem.

Valuable as the above theoretical and conceptual framework may be for structuring the analysis, it is still unknown to what extent the various options to relieve population pressure on land use have played a role in reality in the various tropical regions. In addition it is interesting to gain additional insight as to what extent adjustment patterns differ among the various tropical regions. Therefore an empirical study has been carried out (Hoogeveen, 1992) to try to determine to what extent land in- and extensification and/or changes in land productivity have resulted from increasing population.

The outcome of this empirical work should be viewed against the background of the continuous expansion of the area of land for agricultural purposes in the various tropical countries under consideration. With hardly any exception (Myanmar) the agricultural land area in the tropical countries showed an increasing trend in the period from the mid-sixties on. The data in Table 5.2 clearly illustrate this as well as the diversions between the various regions and countries. The expansion of the land area under cultivation is particularly strong in Latin America, but also significant in the various parts of Asia and in some densely populated countries in Africa.

By definition the following identity holds:

$$Y/L = Y/A * A/L$$

Table 5.2 Agricultural land in 1000 ha 1965–89; average annual growth

	1965–89		1975–89		1985–89	
	%	1000 ha	%	1000 ha	%	1000 ha
Latin America	1.02	3471	0.80	2917	0.62	2373
Bolivia	0.04	13	-0.07	-22	-0.06	-18
Brazil	1.36	2879	1.03	2374	0.80	1959
Colombia	0.63	266	0.62	269	0.49	222
Ecuador	2.12	129	2.69	173	1.55	117
Peru	0.07	22	0.12	38	0.02	7
Venezuela	0.84	163	0.41	85	0.40	86
Africa	0.26	519	0.19	403	0.11	234
Cameroon	0.21	31	0.30	45	0.09	14
Côte d'Ivoire	0.70	43	0.88	55	0.76	50
Ghana	0.10	6	0.03	2	0.04	3
Kenya	0.36	21	0.15	9	0.16	10
Madagascar	0.10	37	0.06	24	0.03	12
Nigeria	0.30	154	0.22	113	0.14	73
Senegal	0.22	24	0.15	16	0.00	1
Uganda	0.91	95	0.84	93	0.23	26
Tanzania	0.20	79	0.04	16	0.04	15
Zaire	0.13	30	0.13	30	0.13	31
Asia	0.76	547	0.58	436	0.38	301
Indonesia	0.40	126	0.23	73	0.22	73
Malaysia	0.42	20	0.42	20	0.42	21
Philippines	0.81	68	0.87	76	0.52	48
Thailand	2.01	333	1.44	272	0.82	169
Myanmar	-0.04	-4	-0.05	-5	-0.08	-8
Papua New Guinea	0.94	4	0.22	1	0.05	0

Source: author's calculations based on FAO Yearbooks.

where:

Y = agricultural production;
L = agricultural labour; and
A = agricultural land area.

A/L therefore is the *land-labour ratio* or a measure of the land extensification of agriculture.

In line with the well-known *Hayami/Ruttan method* (Hayami and Ruttan, 1985), the above identity has for our purposes been rewritten:

$$Log(Y/L) = Log(Y/A) + Log(A/L)$$

If one assumes that Y/L shows a long-term upward trend – this assumption seems realistic in view of the past evidence in most of the tropical countries during the last two or three decades as is illustrated in Table 5.3. – in theory the following three extreme cases could be distinguished:

1) Both Y/A and A/L increase. In other words land productivity increases, but also the acreage per labourer.

2) The *agricultural land productivity*, Y/A, increases so strongly that the amount of hectares cultivated per average labourer, A/L, shrinks.
3) One turns into an agricultural system with a land productivity which is so much lower than before, that the increased labour productivity is to be fully derived from the increased acreage per labourer: Y/A declines whereas A/L increases strongly.

Table 5.3 Average annual growth rates Y/L 1965–1989 (%)

	1965–89	1975–89	1985–89
Latin America	3.23	3.24	2.54
Bolivia	3.13	0.89	0.22
Brazil	3.71	4.08	2.89
Colombia	2.59	1.96	3.48
Ecuador	3.09	3.28	4.20
Peru	−0.9	−0.55	1.43
Venezuela	3.26	1.93	−0.55
Africa	0.21	0.32	−0.01
Cameroon	2.20	1.81	1.15
Côte d'Ivoire	2.75	2.06	1.20
Ghana	1.09	1.09	1.67
Kenya	0.38	0.61	2.25
Madagascar	1.07	0.06	0.18
Nigeria	−0.24	0.51	−1.03
Senegal	−1.35	−2.02	−0.02
Uganda	−1.64	−1.97	−1.21
Tanzania	−0.70	0.40	0.87
Zaire	1.32	0.78	−0.21
Asia	2.48	2.48	1.56
Indonesia	2.70	3.06	2.31
Malaysia	4.45	4.32	5.18
Philippines	1.56	0.71	1.22
Thailand	2.22	1.83	0.25
Myanmar	2.34	3.16	−0.02
Papua New Guinea	2.18	1.61	1.10

Source: Based on author's calculations.

What case materializes in a specific situation can be extremely important from the point of view of tropical deforestation. In case 1 an increasing agricultural population will achieve a higher level of agricultural production both through an intensified use of the land area, but also by extending the acreage per labourer. The need for additional land can therefore be attributed to two factors: the increasing agricultural population as such and the increased amount of land cultivated per labourer. An increased need for land can only be prevented if one could somehow succeed in reducing the number of labourers in agriculture.

In case 2 one succeeds in reducing the land area occupied per labourer through a strong increase in the land productivity. The possibility exists that an increasing population can be fed from the same or even smaller amount of land as before. Deforestation could therefore be prevented.

In case 3 one shifts towards a low land productivity type of agriculture, thus expanding the amount of land needed substantially. This case obviously is the most threatening for the tropical forests.

To arrive at the various trends in both land and labour productivity in the various tropical countries under consideration, data were collected for the period 1965–1989. In order to arrive at comparable estimates of agricultural productivity data were derived from one single source (FAO, 1985 and FAO Yearbooks), and were based on purchasing power parities (PPPs) rather than official exchange rates (van Ark, 1991). Thus in the following:

Y = final output in international dollars 1975 agricultural
    production index, where final output is defined as gross
    output minus feed and seed;
L = economically active population in agriculture; and
A = agricultural land in hectares.

First, the average 1965–1989 agricultural land productivity and its growth rates have been estimated for the three major tropical regions and for 22 countries within these regions.

In Table 5.4. the average productivity figures for the Latin American, African and Asian region are presented.

Table 5.4 Labour and land productivity 1965–89

Year	Latin America		Africa		Asia	
	Y/L	Y/A	Y/L	Y/A	Y/L	Y/A
1965	947	61	291	62	283	238
1966	1005	63	299	65	304	259
1967	1036	65	295	66	291	251
1968	1030	65	288	66	304	267
1969	1072	69	303	72	310	276
1970	1081	70	299	74	319	290
1971	1138	73	292	73	323	296
1972	1166	74	288	74	319	290
1973	1169	74	284	74	346	313
1974	1288	80	292	77	354	316
1975	1300	80	293	79	361	326
1976	1313	80	292	80	367	335
1977	1419	86	286	80	370	339
1978	1407	84	285	81	384	352
1979	1466	86	282	82	390	360
1980	1541	90	284	84	411	381
1981	1627	94	291	88	427	397
1982	1645	95	294	91	433	403
1983	1617	93	284	89	444	413
1984	1691	97	292	93	462	429
1985	1836	105	306	99	479	449
1986	1740	99	313	103	483	454
1987	1904	107	302	102	479	453
1988	1997	112	304	104	497	472
1989	2030	113	306	107	509	487

Source: Based on author's calculations.

From Table 5.4 remarkable differences between regions in average labour as well as land productivity come to the fore. Whereas average Latin American *labour productivity* was much higher during the entire period than in Africa and Asia, the average land productivity in Asia was much higher than in Latin America and Africa where it was more or less the same. Average Latin American labour productivity was not only higher than in Africa and Asia but grew faster as well (3.23 per cent over the entire period as compared to Africa 0.21 per cent and Asia 2.48 per cent).

Finally the growth rates of land productivity have been determined for the various countries, as is shown in Table 5.5. This shows that average Latin American land productivity grew slower than in Asia (2.61 per cent as compared to 3.03 per cent) with the exception of Bolivia with a 3.61 per cent growth rate. Average Latin American land productivity growth was higher than in Africa (2.32 per cent), with some Latin American countries above and some below the African average.

An approach based on the same identity as stated above was subsequently applied at the country level, notably in Brazil, Zaire and Indonesia. This approach tried to analyse to what extent long-term patterns in labour and land productivity in some of the main tropical countries may have differed in the past (1965–1989), and what the

Table 5.5 Average annual growth rates Y/A 1965–1989 (%)

	1965–89	1975–89	1985–89
Latin America	2.61	2.46	1.86
Bolivia	3.61	2.37	2.02
Brazil	2.55	2.62	1.59
Colombia	2.56	2.06	3.26
Ecuador	1.45	0.88	3.08
Peru	1.41	0.84	2.79
Venezuela	2.11	0.99	-1.68
Africa	2.32	2.23	1.92
Cameroon	2.56	1.74	1.55
Côte d'Ivoire	3.53	2.21	1.97
Ghana	2.48	2.66	3.14
Kenya	3.48	3.51	4.93
Madagascar	2.23	1.80	1.91
Nigeria	2.31	2.68	1.00
Senegal	1.08	-0.90	1.63
Uganda	0.22	-0.52	0.78
Tanzania	2.50	2.77	3.12
Zaire	2.02	2.00	1.28
Asia	3.03	2.91	2.07
Indonesia	3.45	3.58	2.82
Malaysia	4.98	4.32	4.79
Philippines	2.40	1.33	2.16
Thailand	2.06	1.83	0.53
Myanmar	3.09	3.98	0.82
Papua New Guinea	2.11	2.28	1.46

Source: Based on author's calculations.

impact of this may have been on the land intensity of agriculture in the three countries. One should again bear in mind the figures represented in Table 5.2, showing an average annual increase of the agricultural land area during 1965-1989, varying from 1.36 per cent in Brazil and 0.40 per cent in Indonesia, to only 0.13 per cent in Zaire. Note in addition that in all three countries the agricultural labour force in 1989 was larger than in 1965 (Hoogeveen 1992, Appendix 3). So, increases in land-labour ratios are due to the expansion of the agricultural area and not, for example, to movements of the agricultural labour force into other sectors.

Figure 5.2 represents the average productivity rates and productivity growth paths for Brazil, Zaire and Indonesia. Since both axes are in a logarithmic scale, lines along which land-labour ratios are the same *(uni-A/L lines)* could be drawn. This was done for three levels of acreage per average labourer: 1, 3 and 16 ha per labourer respectively. In this way the wide variety in the intensity of land use among the three countries could be illustrated. Apparently the amount of hectares per labourer in Brazil is usually some 15 times larger than in Indonesia, and about 5 times as large as in Zaire! Even more interesting are the trends in land intensity in agriculture in these countries. Whereas in Brazil, where agriculture is already land extensive, land extensity in the course of the period increased even further (illustrated by arrow bowing downward), the opposite pattern was shown in Zaire and, although to a much lesser extent, also in Indonesia (arrow bowing upward). (For more data on the three countries, see also pp.102-109.)

Since the most important cause of deforestation in the humid tropics is the conversion of forests into agricultural land, and since for a given demand for agricultural output the demand for additional land is inter

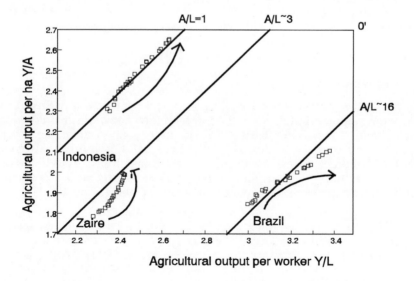

Figure 5.2 Labour and land productivity 1965-1989 in Brazil, Zaire and Indonesia

alia dependent on land productivity in the existing agricultural area, land productivity levels and growth on those existing lands are and will be of utmost importance to the extent of deforestation rates.

> The prime reason for deforestation is the need to break new agricultural land in order to satisfy food requirements of expanding populations. As long as this situation prevails, no forestry projects, however many and large they may be, no forestry legislation, no land use planning and no education will make any significant difference. Deforestation will continue at present shocking rates. The need for food will make that a necessity . . . the main solution to deforestation is to be found in the agricultural sector.
>
> (Birgegard, 1991, p.35)

DEFORESTATION FOR AGRICULTURAL PURPOSES

Introduction

One of the main conclusions of the former sections was that an increasing population will commonly require more land for agriculture, unless the productivity per unit of land is sufficiently increased. Experience in the tropical regions during the last few decades has shown that the area of land under cultivation has risen in virtually all tropical countries, and in some of them significantly so. At the same time empirical data illustrated a wide variety of land-labour ratios among the various tropical countries. Surprisingly enough the data showed an increasing amount of land per person employed in agriculture in those countries which had already a rather extensive way of carrying out agriculture. The latter applies particularly to Latin America.

Here and later on in this chapter we will focus more explicitly on the two factors into which the increment of agricultural production can be decomposed: the expansion (or contraction) of the area under cultivation and the agricultural productivity.

The agricultural frontier

In the humid tropics, increased agricultural production to sustain a growing population has long been obtained by the continual creation of a production frontier. Farmers, particularly young farmers and their families, have moved outwards, seeking new land both for subsistence and commerce, often combining their agricultural activity with hunting, gathering and timber cutting and frequently acting as agents of political expansion. Manshard and Morgan (1988) argue that:

> despite many early attempts to intensify agricultural production, this frontier has continued to exist wherever fresh land has been available, often encouraged by extension of the transport network and the lowering of transport costs. There are still extensive areas of relatively empty land in the humid tropics, debatably capable of sustaining agricultural production but whose clearance must reduce the forest resource and involve serious ecological issues.

As early as 1826 it was suggested by Von Thunen, in his model of land-use rings around an isolated market, that whilst new developments in agriculture would tend to appear near the centre of the system, simpler forms of subsistence or near-subsistence production, together with the commercial production of certain crops and livestock, would continue on the periphery. In the broad sense in which one may accept that there are certain parallels between Von Thunen's model and a world system of agriculture, that is generally true today.

(Manshard and Morgan, 1988)

Both authors continue by arguing that the production frontier has increasingly become broken into distinct zones of settlement expansion, each with its own distinctive characteristics.

These are not only from environmental differences within the humid tropics – as, for example, between marginal lands in great river basins such as the Amazon and mountain plateau fringe zones as in Thailand – but also from differences in the political and production base or core area of each expansion, together with differences in society and economy, which have led to different objectives and the emergence of many different kinds of pioneer settlers, despite the existence of some common features and trends. Nor should one forget that these relatively empty lands frequently have host populations with extremely varied attitudes to the colonization in their midst.

In addition, the increasing importance of accessibility to commercial and political core areas in the development of the peripheral lands is a major feature of the changing nature of the production frontier. Road building not only opens peripheral regions to planned settlement expansion with commercial opportunity but also creates possibilities for spontaneous settlement, sometimes impossible to control and with unexpected results.

The identification of a production frontier, of a zone of colonization and pioneer settlement is not always clear. Not all new settlements are spatially peripheral. There are 'hollow frontiers', inner zones of relatively empty or weakly developed land. The distinction often suggested between higher intensity systems towards the core area and lower intensity towards the periphery does not always appear. Whilst rubber production in the Amazon region was the product of an extensive gathering economy, rubber production in Malaysia appeared in a rainforest frontier but was associated with an intensive high-level technology.

The picture presented is of a highly complex pattern of developments, somewhat removed from the earlier conceptualization or modelling of the production as a zone of pioneer subsistence farmers with large-scale commercial systems, ranches, and plantations in the more accessible areas, all following in the wake of primitive hunting and gathering economies, occasionally also with a commercial character. Even the distinction between spontaneous and planned forms of settlement is not equally clear in every region.

In reality, pioneer settlements are mostly neither so spontaneous, nor so planned. The planners are not just government or government agencies such as the army or the departments of forestry or agriculture but private corporations, including multinational companies, international and national finance agencies, estate owners, and land speculators. The settlers are not just small groups of young farmers looking for larger farms and fresh land, or groups of shifting cultivators like the hill tribes of

Malaysia, but also farmers pushed out of their former holdings by the development of large-scale, mechanized, commercial agriculture; farmers looking for land on which to develop their own commercial mechanized or plantation production; long-distance migratory farmers who specialize in opening up new lands and who are forever moving on; short-distance migrants expanding village lands; commuter farmers moving daily to a local periphery; farmers responding to inducements offered to settle on the frontier; war veterans; and even urban-based labourers and others returning to farming or miners and timber cutters engaged in spare-time cultivation. They may farm as individuals, they may be under contract to dealers, or they may be in co-operatives or grouped as estate tenants.

A great deal of frontier settlement is multi-purpose. Spontaneous settlers may have more than one objective, such as both subsistence and commercial cultivation, combined with gathering or timber cutting and the hope of rising land values and the eventual sale of property. Planned settlements may have political or military objectives and may be intended also to relieve population pressure on land or on urban jobs and resources.

(Manshard and Morgan, 1988)

Some differences among the major tropical countries

Figure 5.2 clearly showed, with the help of some trends on labour and land productivity of Brazil, Zaire and Indonesia, to what extent agricultural developments differ between these tropical countries. This suggests that patterns of change of the agricultural frontier differ markedly among the major tropical regions of the world. Indeed, most observers of the forest situation in Latin America agree that cattle ranching in that country is the primary cause of deforestation: this is in direct contrast with tropical Asia and Africa where shifting cultivation and logging are considered the most important causes. Some additional data may serve to illustrate the variety in scope of the deforestation problem in the major tropical countries.

Brazil

In Brazil population has grown from a total of 17 million in 1900 to over 150 million in 1990. This corresponds with an average annual rate of growth of 2.8 per cent, and reflects the country's traditional point of view that a large population enhances its economic and political position in the world. This rapid expansion of the population size was reflected in a strong need for additional agricultural land. Although the colonist small farmers have contributed to the deforestation process in Amazonia, it is commonly acknowledged that by far the major part of the land that ends up cleared of forest (some 90 per cent according to some sources) can be attributed to cattle pastures of the ranchers. Indeed, the increased land area adopted for permanent pasture and cattle ranching is typical for Latin America, and for Brazil, Colombia, Ecuador, Peru, and Venezuela in particular. In fact the amount of hectares used for permanent pasture between 1985–1987 and 1975–1977 only increased in Latin America; it stabilized in Africa and somewhat decreased in Asia. This is illustrated in Table 5.6.

Table 5.6 Land converted to permanent pasture

	1000 ha 1985–87	*% change since 1975–77*
Latin America	284,374	6.4
Bolivia	26,800	-1.2
Brazil	167,000	6.4
Colombia	39,804	7.3
Ecuador	4900	61.5
Guyana	1230	17.0
Peru	27,120	0.0
Surinam	20	22.9
Venezuela	17,500	3.4
Africa	182,311	0.0
Angola	29,000	0.0
Benin	442	0.0
Cameroon	8300	0.0
Centr. Afr. Rep.	3000	0.0
Congo	10,000	0.0
Côte d'Ivoire	3000	0.0
Eq. Guinea	104	0.0
Gabon	4700	-1.4
Ghana	3410	-2.8
Guinea	3000	0.0
Guinea-Buisseau	1080	0.0
Kenya	3740	-1.0
Liberia	240	0.0
Madagascar	34,000	0.0
Nigeria	20,970	0.8
Senegal	5700	0.0
Sierra Leone	2204	0.0
Togo	200	0.0
Uganda	5000	0.0
Tanzania	35,000	0.0
Zaire	9221	0.0
Asia	488,894	-1.1
Indonesia	44,000	0.0
Malaysia	123,338	-5.0
Philippines	1180	26.2
Thailand	740	32.1
Kampuchea	50	0.0
Lao Dem. Rep.	10	0.0
Vietnam	310	14.0
Myanmar	319,080	0.0
Fiji	60	-7.7
Papua New Guinea	87	-18.5
Solomon Isl.	39	0.0

Source: WRI (1990), Table 17.1, p.268.

These pastures became particularly popular from the 1960s on, partly through the growing network of paved inroads, e.g. for logging or petroleum and mineral exploration, financed by the federal government and the multinational development banks. Other stimuli were the various fiscal government measures and their practice of 'first come, first served titling', by indirect subsidies through government subsidized research and the free provision of technical assistance; and by speculation. To give an example, according to WRI (1990) some 530 livestock investors benefited by $730m in tax exemptions and subsidized rural credit between 1966 and 1983. Estimates of others point at even higher figures. Pearce et al. (1990) argue that over the period 1965–1983 direct tax credit subsidies worth $1.4 bn were granted to 808 existing and new private investment projects, 42 per cent of which went to 469 livestock projects, virtually all for beef cattle production. Repetto (1990) argues that the subsidized long-term loans, tax credits, tax holidays and write-offs to the existing (some 600) ranches in the Brazilian Amazon covering some 12m hectares – roughly four times the surface of the Netherlands – have already cost the treasury more than $2.5 bn in lost revenue.

Forest is converted to pasture lands in a number of various ways (Shane, 1986). Sometimes peasants cut and burn the vegetation for planting their subsistence crops such as corn, manioc, yucca and beans, but leave the site because of the declining soil fertility after two to three harvests. The land can then subsequently be taken over by cattle interests. In other cases cattle ranchers are employing colonists or using heavy machinery to clear-cut large tracts of virgin forest. In some instances cattle ranchers finance farmers to clear land for eventual conversion to pasture.

But the most frequent method used to remove vegetation is for the ranchers to employ from several to scores of workers who use machetes and chain saws to cut the forest. Once the cut or defoliated forest vegetation has dried, it is burned, which serves not only to remove the debris but to release stored nutrients into the soil. After the forest has been burned, the ash-strewn land is sown with any of a variety of exotic grasses, most of which originated in Africa.

Box 4 Cattle ranching: the controversy

There is a fierce controversy between experts as to whether cattle ranching is a viable development activity in the humid tropics. Illustrative in this respect is the following case derived from Shane, 1986.

Fueling the controversy are differing reports concerning the 72,000 hectare King Ranch operation, Fazenda Uraem, in Para, Brazil. According to a King Ranch spokesman in Kingsville, Texas, the operation is a success. Fazenda Uraem, which is owned by a corporation that includes the King Ranch, the Swift-Armour meat Packing Company, and a consortium of Brazilian interests, was started in 1972 and, according to the spokesman, has experienced no problems with soil compaction or deterioration. Toxic weeds are controlled with the herbicide *Tordone* and by burning. But other agronomists who have visited Fazenda Uraem are less enthusiastic about the operation's success than the King Ranch spokesman. The independent scientists say that due to the usual drop in the soil's nutrient levels, pastures are being lost to weed and brush invasion after five years while herd mortality due to toxic weeds is as high as five to ten percent, and that diseases claim others whose resistance has been lowered by nutritional deficiency.

It is commonly acknowledged that the above-mentioned huge cattle ranches with an average size of over 20,000 hectares are in fact uneconomic, and can only survive through the subsidies and the favourable tax regime on the one hand and through overgrazing on the other. Pearce et al. (1990) argue, for instance, with respect to the Eastern Amazon that without overgrazing, real land values must appreciate at the rate of 30 per cent before the investments become economically viable, unless, of course, they receive subsidies. Surveys with respect to cattle ranching in the Brazilian Amazon area showed that meat output averaged only 9 per cent of what was projected and that many ranches were reorganized and resold repeatedly, having served only as tax shelters.

All this, and the rapid increase in the demand for beef in the OECD countries – between 1958 and 1980 USA beef imports quadrupled! – contributed to the fact that Brazil has the second largest cattle population in the world, and the size of its permanent pasture keeps expanding: it has increased by some 6.4 per cent since 1977, corresponding to some 10 m hectares. The expansion of the area for permanent pasture in the rest of Latin America during the same period was in the same order of magnitude. In the whole of the Amazon basin there may be as many as 100,000 ranches, the vast majority of which have occurred by converting forests to pasture lands. This expansion largely contributed to the process of tropical deforestation. One estimates that over the past two decades more than 20 m hectares (3 per cent) of humid tropical forest in Latin America have been converted to cattle pastures. At least one half of this conversion has taken place in Brazil's Amazon, a quarter in Mexico, and the rest in Colombia, Peru, Venezuela and Central America.

Experience has shown that the exploitation of the land for permanent pasture often led to a rapid process of land degradation. After the forests were cut and burned and planted with forage grasses – during the first two or three years growing well because of the phosphorous and other nutrients released by burning the rainforest – these grasses then became gradually overtaken by shrubs and non-forage grasses, probably due to the decreasing availability of phosphorous, aggravated by overgrazing (often one animal unit per hectare). Sometimes degraded pastures were temporarily recuperated by weeding and sprouting, or even by scraping the soil surface with bulldozers, fertilizing and replanting with forage grasses, but this eventually reinforced the initial process of land degradation and environmental decline. Moreover, the potential for regeneration was adversely affected. To give an example (WRI, 1990), a pasture in Venezuelan Amazon that was repeatedly burned to maintain the grasses had regenerated only 50 per cent of its biomass after 80 years of abandonment. Another example, eight years after abandonment, a pasture near Paragonimus that had been bulldozed was dominated by weedy grasses and shrubs, noxious insects, and supported no tree species native to the original forest.

Even though cattle projects in the humid tropics are no longer pushed by, for instance, the World Bank and other official agencies since the end of the 1970s, and even though the Brazilian government no longer subsidizes pasture formation in Amazonia (in fact according to Millikan, (p.131), as much as 90 per cent of Amazonian pasture land may have been formed without government subsidies) 'pastures continue to be formed as a means of securing and retaining title to land, thereby capturing escalating land values. Cattle are valued as an inflation-free investment and as a source of milk products and status, even when cattle production is not profitable' (WRI, 1990, p.108). It has even been suggested that the mystique in Latin America of what some call 'the cult of the bull', is a social factor which, to some Latins, is almost as important as the economic aspects of cattle ranching.

All the above factors contributed to the rapid deforestation processes in the country, estimated at over 10 m hectares of the humid tropical forest over the past two decades. Once this process has got started deforestation can accelerate by the increased threat of fire. Indeed, primary forest cannot be ignited even after months of drought whereas during the dry season, degraded pastures can burn within one day of rainstorm. This explains why such extensive forest fires can take place in Amazonia. According to Seltzer et al. of the INPE, the national Space Research Institute of Brazil and based on satellite imagery, in 1987 not only eight million of virgin forest of the total five million km^2 of Brazil's legal Amazon area were cleared, but also 20 million hectares in Amazonia burned, a large portion of which had supported forests exploited for timber.

Despite the significance of the cattle ranching problem in Brazil, the adverse impact of shifting cultivation on the tropical forests should not be ignored. These activities can also be held responsible for part of the deforestation. To combat deforestation in Brazil by decreasing the demand for forest lands from those shifting cultivators will, again, not be an easy task however. As in the other tropical regions, the reasons why they head for the forests – population growth, maldistribution of established farmlands, lack of agrotechnologies for intensive cultivation, and inadequate rural development – generally require major and comprehensive development efforts: 'the pressures that have been driving the colonists into Rondonia, Acre and other settlement areas . . . show no sign of abating within the foreseeable future' (Myers, 1991).

Some government measures geared to the smallholders probably have even aggravated the deforestation problems. The government has, since 1970, subsidized small-farm settlement in Amazonia (ranging from $3900 to $13,000 per family). Underlying this policy is the government target that smallholder settlements should eventually extend across a total of between 200,000 km^2 and 430,000 km^2. The settlement programmes thus far have had only a limited impact in terms of relieving overall population pressure, but a serious impact in terms of environmental decline. The two largest programmes in Rondonia, for example, were by 1980 responsible for an estimated 6.6 per cent of the forest area altered in Amazonia. At the same time, through these programmes at that time only 48,817 families have been given title to land (Pearce et al., 1990). So, the impact on reducing rural population pressure in the rest of Brazil has been largely superficial.

In the meantime there is evidence that there is indeed a positive correlation between the consumption of smallholder subsidies and the area of forest cleared. Moreover, the experience has shown that forest is not only cleared for the initial establishment of the smallholder settlements. Settlers tend to abandon their designated sites after just a few years to take up a lifestyle of slash-and-burn cultivation or '*shifting ranching*', that way further aggravating the problem.

Indonesia

In Indonesia the picture is quite different. A recent study carried out under World Bank auspices (World Bank, 1990, p.3) concludes that some 500,000 of the 900,000 hectares deforested per annum should be attributed to smallholder conversion, while development projects including transmigration programmes can be held responsible for another 250,000 hectares. The remainder was considered to be due to logging, 80,000 hectares, and fire losses, 70,000 hectares. Clearing of the forest in present day Southeast Asia is therefore first of all linked to the rapid expansion of pioneer settlement, either by smallholder pioneers or by agricultural entrepreneurs growing crops for vital export earnings.

Due to the combination of a number of factors, official transmigration had virtually halted in 1986. These factors included:

- the high costs of financing the Indonesian *transmigration programme* (an estimated $9000 per family in remote areas and additional expenditure on rehabiliting poor soils);
- the increasing criticism of environmental groups;
- the subsequent reluctance of donors to support the programme; and
- a lack of suitable sites.

Nevertheless, since 1979 around 540,000 families have been resettled in the Outer Islands. By the end of the 1980s an estimated one million hectares of forested land have been converted to agricultural land for transmigration (Pearce et al., 1990).

However, the ending of official transmigration has not stopped spontaneous migration to the Outer Islands. For every family officially resettled, an additional family probably migrated to the Outer Islands spontaneously (ibid). Spontaneous migrants receive hardly any support for settling in the Outer Islands; consequently, they are more likely than sponsored transmigrants to depend on converting new areas of forests. Moreover, their unfamiliarity with Outer Island forest systems may lead them to develop less sustainable and more destructive shifting cultivation-based systems. The spontaneous opening up of new agricultural regions often passes nearly unnoticed by official reports, statistics or research:

It has grown to such a scale and importance that it outstrips the official schemes in area, in population, and also in productivity. One of the consequences of the spontaneous land colonization is that the process has become rather unmanageable, and increasingly creates conflicts either between the forest authorities and the spontaneous settlers, or even between the forestry administration and the state directed land clearing schemes. Ethnic and cultural pluralities tend to aggravate any frictions on land.

(Uhlig, 1988)

Tree crop development on the Outer Islands, another form of officially encouraged colonization, is also contributing to deforestation. Due to the financial constraints facing the government of Indonesia, the rate may have slowed from an initial 40,000 to 20,000 hectares per annum since 1986 (Pearce et al., 1990). The establishment of palm oil estates, however, continues to be a major priority. In addition to benefiting from generous subsidies on fertilizers and other inputs, pricing policy interventions have been used to encourage substitution of palm oil for coconut oil on the domestic market. Export restrictions, and in some years export taxes, have also been applied to ensure adequate domestic supplies. To meet this induced domestic demand and provide a surplus for export, oil palm production has been rapidly expanded. Between 1979 and 1984 the area under palm oil grew at an annual rate of 9 per cent. The major assumption behind this policy is that forest lands on the Outer Islands can be relatively easily and inexpensively converted into palm oil estates. Around 78 per cent of Indonesian palm oil production currently comes from Northern Sumatra, and around 70 per cent of the total area planted is on state-owned plantations. Recently, however, the Indonesian government has actively encouraged a system of nucleus estates on the Outer Islands, where a government-owned estate company finances investment for smallholder oil palm planters. The rapid expansion of the total area of production through the nucleus estates scheme is seen as essential to expanding future palm oil production, but the scheme has been afflicted by problems with over-fertilization, pest and disease attacks, and the poor quality of converted forest soils (Pearce et al., 1990).

Zaire

Deforestation rates in Zaire are still relatively low (see also Table 1.1a), despite the fact that the area under cultivation for tree crop plantations gradually expands, partly based on slightly better world market prices for some of Zaire's major cash crops for export, coffee and palm oil. The major threat to the tropical forest, however, comes from shifting cultivation. Slash-and-burn cultivators are encroaching upon the country's forests in increasing numbers. The predominantly rural population, growing at a rate of some 3 per cent per year, is generally involved in underdeveloped agriculture (which comes to the fore, among others, in low agricultural productivity levels – see also Figure 5.2) largely of a slash-and-burn type. The number of slash-and-burn cultivators is increasing rapidly not only due to the rapidly expanding population, but also through the greater marginalization of the country's poorest people after the decline of the national economy due to the collapsing prices of copper, the country's main export commodity.

These shifting cultivators also create a threat to the Zairian rainforests because of their ill-adapted farming techniques:

> Over the last few decades the Ituri forest has become a settlement frontier into which families from the rapidly growing, densely populated savannas and highlands from the North and East are immigrating in ever-increasing numbers. These immigrants often bring with them farming techniques that, although applicable to the fertile volcanic soils of the eastern highlands, are incompatible with sustained exploitation of the forest and result in the establishment of short-duration fallows, which are rapidly exhausting resources.
>
> (Wilkie, 1988, p.126)

These destructive farming methods are also pointed out by Myers (1991) where he argues that 'while the total of forestland farmers is still low in proportion to the forest estate, they affect three times as much forest each year as do their counterparts in West Africa'. So, even although the Zairian rain forests are at present relatively unaffected, future encroachment is likely to increase unless the high population growth rate slows down and the economic situation improves.

One major impression gained from the above is that landownership generally plays a major role with respect to land use patterns and thus to deforestation. The common picture in the tropical areas is that the wealthy hold most of the cropland. In Latin America, for example, 93 per cent of the farmland is in the hands of just 7 per cent of the people; in Java 1 per cent of the farmers owns a third of it, while for instance in Kenya, the new African elite holds a higher proportion of the land than did the British imperialists (Manolas, 1991). Various authors claim that land reform in the older agricultural areas would relieve the pressure on the forests by shifting cultivators; the more so because much of the good land held by the big landowners is farmed inefficiently. Land reform, however, is withheld in many countries for internal political reasons. Instead, governments sometimes prefer to relieve the pressure for land reform by state sponsored colonization schemes for the rainforests. Standing (1984) even suggests that 'colonisation may be expected to be amply tried out in a country where the landowners are powerful and where virgin, state-owned lands are plentiful'. He mentioned in this respect: Brazil, Chile, Colombia, Guatemala, Peru, Venezuela, India, Malaysia, The Philippines and Indonesia.

CONSTRAINTS TO AGRICULTURAL INTENSIFICATION

Various institutional constraints will contribute to retarding agricultural change. It is clear that all these constraints interact to cause a low production, therefore improvement in one of those factors will not necessarily guarantee an increase in production. One of the main institutional factors preventing easy adjustments in shifting agriculture to changed circumstances is the lack of sufficient transport possibilities in areas with shifting cultivation. These difficulties restrict the supply of inputs for the farming system and the transport of agricultural outputs. The relatively high transport costs compared to less isolated areas cause unfavourable price ratios of inputs and outputs of the farming system.

Institutional provisions such as information and research institutes, credit and marketing facilities etc., are also highly dependent on transport. Unfortunately, government usually pays little attention to less productive areas. Furthermore, as people in these areas lack political influence, the government spends its scarce means in more developed areas, thereby reducing further the development possibilities of shifting cultivation areas that are already behind.

Another important constraint with respect to increased productivity is the lack of investments. Although there has been a shift from communal to private tenure in shifting cultivation areas, there are still areas where communal tenure prevails. In those areas, uncertainty about property or the continuous use of land is a serious constraint to investments in capital and/or labour.

But even if individual tenure exists, investments are minimum because of the low incomes and low incentives to invest. Generally speaking, all the factors that contribute to low production also restrict the opportunities to invest. Credit facilities are too poor to solve the low investment problem, especially for small farmers. Besides that, credit is mainly given for the purchase of seeds, fertilizer etc.

Especially in humid climates, where shifting cultivation is practised without machinery and fertilizer, rented labour is the most important input. This means that rented labour has to be paid by the farmers themselves, which is generally impossible because of their lack of cash flow.

In many developing countries, government aims at low food prices for the urban population; low prices for agricultural products result from this policy. Low prices are also a constraint to the increase of agricultural production, despite the need for cash income. Farmers are inclined to look for supplementary off-farm jobs for cash income, while their agricultural production is mainly for subsistence. Especially in areas where shifting cultivation is practised, the input-output price ratio is too unfavourable to have any economic incentives to raise production.

Another government policy that can have a negative impact on increased production is tax policy. In developing countries one often sees that a relatively heavier tax burden is put on the agricultural sector. Besides resulting in decreasing incentives to raise production, the cash flow position of the farmers worsens.

In the forest regions and in the tsetse-infested savannas, there is a lack of livestock to provide animal manure. Mineral fertilizer is usually not worth while, because of unfavourable cost-return ratios, which can hardly be avoided in shifting systems.

Differences in yield between the tropical countries, as came to the fore in the productivity comparisons on pp.96–100, and between those countries and more developed nations, are partly caused by differences in agricultural inputs. The differences are still substantial if most of the tropical regions are compared with the western countries. European agriculture (WRI, 1990, p.277), for example, applies two and a half times more fertilizer per hectare of cropland than the global average, while South America uses one-third and Africa one-fifth of the world average. Other inputs show similar distributions. Two-thirds of the world's agricultural machinery is concentrated in North America and Europe, which farm one-third of the world's cropland. Africa, in contrast, with 10 per cent of the world's crop area, owns only 2 per cent of the world's tractors, many of them out of operation because of high fuel prices and a lack of spare parts.

A study by Hoogeveen (1989) analysing the relation between differences in land productivity and, among others, differences in the use of agricultural inputs, concluded that the difference between 1985–1986 land productivity in the Benelux and the tropical regions could, for 13 per cent, be attributed to differences in fertilizer use.

The international data on the use of some technical inputs in tropical regions and countries are shown in Table 5.7. The land augmenting factors, especially, such as irrigation and the use of fertilizer and pesticides have a serious impact on land productivity and explain the variety between the countries to a large extent. Illustrative in this respect is the contrast between the high average land productivity level in Asia, as was shown in Table 5.4, and the heavy use of technical inputs in most Asian countries on the one hand and the generally low land productivity and input appliance in Africa on the other.

At this point the question arises what factors prohibit the increase of agricultural production in the established tropical agricultural areas through a more intensive use of technical inputs. This question is especially relevant if one is reminded of the fact that the alternative subsistence-level farming involves agricultural production in the (for farming) less suitable tropical forest environments. Past efforts such as the introduction and support of the application of green revolution technology and other agricultural development programmes have not resulted in a widespread conversion to more productive, and input intensive agricultural systems in the tropics.

In general, the agricultural development programmes emphasized the use of modern technology by providing subsidies for agricultural inputs. These programmes did not succeed in converting subsistence level farming in the tropics to more productive agricultural systems on a large scale. Problems associated with the subsidizing of agricultural inputs by local governments are discussed later on. The essentials of the green revolution were the development of new varieties, especially rice and wheat seeds, which were capable of responding much better to artificial fertilizer than were traditional varieties.

According to Todaro and Stilkind (1981), agricultural development programmes did not reach the majority of peasant farmers. These programmes 'favoured large landowners because only they could afford the new machines, seeds, fertilizers, and the risk associated with a new technology.' Subsistence-level farming persists, however, because

Table 5.7 Technical inputs in tropical regions and countries

	Irrigated land as a % of cropland		Kg fertilizer per ha of cropland		Average annual pesticide use*		Tractors annual		Harvesters annual %	
							Average no.	% change since	Average no.	% change since
	1975–77	1985–87	1975–77	1985–87	1975–77	1982–84	1985–87	1975–77	1985–87	1975–77
Latin America										
Bolivia	4	5	1	2	612	833	790	13	278	42
Brazil	2	3	41	49	59,292	46,698	775,000	148	42,000	31
Colombia	6	9	49	81	19,344	16,100	33,813	36	2417	28
Ecuador	20	21	26	34	5445	3110	8000	52	700	29
Guyana	30	26	26	32	705	658	3560	5	419	4
Peru	35	33	38	43	2370	2753	14,933	19	0	NA
Surinam	79	88	85	178	974	1720	1703	38	134	18
Venezuela	8	9	44	143	6923	8143	44,667	43	4733	92
Africa										
Angola	NA	NA	4	4	NA	NA	10,263	8	0	NA
Benin	0	0	1	6	NA	NA	118	29	0	NA
Cameroon	0	0	3	7	NA	NA	920	197	0	NA
Centr. Afr. Rep.	NA	NA	1	1	NA	NA	185	44	13	60
Congo	0	1	4	5	NA	NA	687	5	45	79
Côte d'Ivoire	1	2	14	9	NA	NA	3350	41	53	112
Eq. Guinea	NA	NA	0	0	NA	NA	99	3	0	NA
Gabon	NA	NA	1	4	NA	NA	1373	31	0	NA
Ghana	0	0	10	0	NA	NA	3800	17	427	110
Guinea	4	4	1	1	NA	NA	180	80	0	NA
Guinea-Buisseau	NA	NA	1	1	NA	NA	48	40	0	NA
Kenya	2	2	22	46	935	1307	8536	41	520	22
Liberia	1	1	15	5	1223	310	318	22	0	NA

	Irrigated land as a % of cropland		Kg fertilizer per ha of cropland		Average annual pesticide use*		Tractors annual		Harvesters annual %	
	1975-77	1985-87	1975-77	1985-87	1975-77	1982-84	Average no. 1985-87	% change since 1975-77	Average no. 1985-87	% change since 1975-77
Madagascar	18	28	3	4	NA	1630	2800	14	138	41
Nigeria	3	3	2	10	NA	4000	10,533	37	0	NA
Senegal	3	3	9	4	NA	NA	467	22	147	24
Sierra Leone	1	2	1	2	NA	NA	483	200	6	240
Togo	0	0	2	7	NA	NA	320	167	0	NA
Uganda	0	0	0	0	NA	23	3800	109	13	41
Tanzania	1	3	6	9	2992	5733	18,550	2	0	NA
Zaire	0	0	2	1	NA	NA	227	52	0	NA
Asia										
Indonesia	26	34	27	100	18,687	16,344	12,511	37	16,471	30
Malaysia	7	8	68	154	NA	9730	11,567	83	0	NA
Philippines	14	18	34	50	3547	4415	19,767	53	587	59
Thailand	15	20	14	26	13,120	22,289	130,333	347	0	NA
Kampuchea	3	3	0	0	1593	833	1358	1	20	0
Lao Dem. Rep.	6	13	0	2	NA	NA	800	100	0	NA
Vietnam	18	28	58	61	1693	883	41,500	344	0	NA
Myanmar	10	11	5	18	3721	15,300	10,282	28	38	84
Fiji	0	0	52	79	NA	NA	4450	31	0	NA
Papua New Guinea	NA	NA	20	30	NA	NA	1163	0	450	38
Solomon Isl.	NA	NA	0	0	NA	NA	NA	NA	NA	NA

* in metric tons of active ingredient

Source: WRI (1990), Table 18.2, p.281.

'farmers cannot afford either the seeds and equipment to plant cash crops or the risk that harvests may fail or prices drop.'

Arguments as to why *green revolution technologies* are hardly adopted in Africa, despite their considerable potential contribution to the development of African agriculture, and why there are few examples of successful African rural development programmes, are given by Levi and Havinden (1982).

> Why is it that alien technologies, despite their apparent promise, have had such a remarkably small impact on African agriculture in general? Part of the answer – possibly the most important part – lies in their very alienness. By definition, the fact that they are alien means that they cannot be supplied from within the rural economy. This implies that unless they are donated *gratis*, and even if they are subsidised, the rural economy must be able to generate an extra cash 'export' to pay for them, or else must be prepared to forgo some of the meagre output it already produces in order to provide the wherewithal for the new technology. Although outside aid has made alien technologies available to some fortunate people, it has never been anywhere near large enough to make them available generally, nor is it likely to be. That being so, there remain the resources of the people themselves, who are mostly very poor. Because of their poverty, any amount of their production foregone, even in exchange for productive inputs, represents a considerable sacrifice: the subjective cost is high. This high cost has to be weighed against the expected returns . . . Because the return is not immediate . . . a family close to starvation will not exchange or sell food for fertilizer even if the return is certain and attractive . . . Simply because of the delay in returns, the poorer the family the less fertilizer will it be willing to buy. Borrowing may be possible, but non-institutional rates of interest tend to be very high, averaging around 50 per cent per annum and often 100 per cent or more on short-term credit. Institutional credit at relatively low rates of interest has been seen as a possible solution, but inevitably it has to be rationed and in practice has to be channelled towards the better-off if a high level of default in repayment is to be avoided.

Some additional arguments why the use of fertilizers in the tropics is not greatly increased, even if without such an increase it is difficult to see how fertility can be maintained and more and better food can be produced for the rapidly increasing populations, were presented much earlier by Webster and Wilson (1971). The arguments put forward more recently by WRI (1990) are along similar lines. According to them it is highly unlikely that the poor performance of agriculture in the tropics can and will rapidly be solved by a massive increase of fertilizers, because using more fertilizers is not likely to be highly effective in most parts of the tropics unless it is accompanied by improvements in husbandry and in soil and water conservation. Indeed, there is a risk that a greater availability of fertilizers might make matters worse by enabling farmers temporarily to make more intensive use of land without proper conservation measures, thus leading to increased denudation and erosion. Furthermore there is a danger that a large use of fertilizers in peasant farming could induce conditions of nutrient imbalances in the soils.

Therefore, in order to increase agricultural production in the tropics significantly, and at the same time to maintain soil fertility, not only an increase in the use of fertilizers, or any other single input, is required, but a more general extension of more productive and fertility-building farming systems. In order to meet the increasing demand for agricultural products to feed their growing populations and to increase their export earnings, many governments in developing countries spend considerable amounts annually on subsidies for agricultural inputs to increase agricultural production. Yet these subsidies have so far not succeeded in the widespread establishment of farming systems and the use of farming techniques, capable of raising production and absorbing excess labour to such an extent that the pressure on marginal lands such as the forests is greatly relieved. Much attention has been paid in the literature to problems associated with subsidies on agricultural inputs as a market distortive instrument.

To give an example, fertilizers have been heavily subsidized since the 1960s. Subsidy rates in the tropics as a percentage of delivered costs commonly range from 50 to 60 per cent, and are sometimes even higher, such as in Nigeria during the 1980s where a 90 per cent rate was reached. The reason why fertilizers (or other agricultural inputs) are subsidized is to introduce the new technology, to induce learning by doing, or to overcome faulty perceptions of risk. However, economic arguments for large and continuing subsidies are shaky. Repetto argues:

> If farmers are slow to adopt chemical fertilizers, it may be because of problems of distribution, extension, or availability of complementary inputs, not because of price. In parts of Africa and other regions of low population density, fallowing might be a more economical approach to restoring soil fertility. Fertilizer subsidies only partially offset explicit and implicit taxes on agricultural output, and are often captured by those who do not really need them (large commercial farmers on irrigated land) and those for whom they are not intended (producers and distributors).
>
> (Repetto, 1989, pp.74–75)

Another drawback of subsidies is that they create market distortions and therefore contribute to the inefficient use of fertilizer and other complementary inputs, resulting in the waste of costly inputs and increased pollution. Fertilizer subsidies worsen land degradation because:

> These subsidies artificially lower the cost of maintaining and restoring soil fertility and so reduce farmers' incentives to practise soil conservation . . .. Specifically, subsidies induce a substitution in favour of chemical fertilizers and against organic manures and crop residues . . .. Organic and chemical fertilizers are not perfect substitutes for one another . . .. Although chemical fertilizers provide cheap concentrated sources of certain nutrients, organic manures also provide a variety of micronutrients and improve soil structure . . .. In addition, when used with chemical fertilizers, organic manures improve yields and offset the sharp decline in marginal returns to chemical fertilizers that most South and Southeast Asian countries have experienced. Heavy fertilizer subsidies have become an enormous fiscal burden with uncertain benefits and substantial environmental costs both on and off the farm.
>
> (Repetto, 1989, pp.75–76)

Another agricultural input heavily subsidized in tropical countries is the use of pesticides, varying as a percentage of full retail costs in 1985 from, for example, Ecuador – 41 per cent; Colombia – 44 per cent; Ghana – 67 per cent; Indonesia – 82 per cent; to Senegal – 89 per cent (ibid.). Just as for fertilizers it is doubtful whether the arguments for large and continuing subsidies hold. Pesticides continue to be subsidized even although the technology is long familiar and the bulk of the subsidies go to large commercial farmers who do not need them. Moreover, subsidies can contribute to the inefficient use of pesticides: 'pesticides should be used only at key stages in the life cycles of pests or crops or when damage to crops reaches a predefined threshold. By lowering pesticide costs to farmers, subsidies artificially depress this threshold and encourage prophylactic applications' (ibid, p.74). Some argue that the resulting excessive use of pesticides in developing countries forms a threat to the health of millions of people and an ecological risk. Subsidies, furthermore, lower the costs of pesticide use relative to other less risky control methods such as planting resistant varieties of crops, destroying infected plants, and altering planting dates. Pesticides subsidies thus can block the development of more sustainable pest management practices.

The fundamental problem of the heavy subsidies on irrigation (at current prices, US$250 billion have already been invested in irrigation in developing countries, and US$100 billion more may be spent this century to create more capacity), lies in the financing system. Revenues in most countries do not even cover operating and maintenance costs. Charges are also small relative to the value of water to farmers, especially during peak periods of need. This financing system thus creates chronic excess demand and huge economic rents for those able to obtain water from public systems. Although the benefits in terms of increased agricultural production have been substantial, the economic and environmental costs have been as well:

> Costs have been much higher and benefits lower than projected when investments were made . . . Impounded water and canals provide breeding grounds and habitat for carriers of malaria and schistosomiasis. They have displaced whole communities and flooded valuable crop and forest lands, threatened critical ecosystems, and wiped out anadromous fish populations. The disruption of river hydrology downstream has caused erosion and sedimentation and had a great impact on estuaries and even deltaic fisheries.
>
> (Repetto, p.76)

In particular the promotion of agricultural mechanization forms a threat to the tropical rainforests if, as is often the case, only the large landowners benefit. Subsidies generally promote mechanization on the larger estates and contribute to the reduction of rural employment. Smallholders not only suffer a competitive disadvantage because their typically fragmented holdings are not suitable for mechanized operations, but also because they are dependent on seasonal wage employment for their income. So they suffer through labour displacement as well. The marginalization of those smallholders causes them to seek a livelihood elsewhere. This often means a shifting cultivator existence in the forests. In addition, agricultural mechanization in the forest regions themselves contributes to the damage to the soil and the loss of biomass. Using heavy equipment instead of traditional methods for land clearance in tropical regions has sometimes devastated the soil. Through these practices nutrients in the biomass have been lost, thin topsoils have been scraped off, the ground has compacted so that water cannot infiltrate, and erosion rates have risen enormously.

Finally one should consider the subsidized agricultural credit programmes. They are at least as widespread in developing countries as subsidized fertilizers and other inputs and are even more questionable on economic grounds. Special loan funds with interest ceilings are often set up for purchasing particular inputs, growing particular crops, acquiring particular assets such as cattle or tractors, and developing land by clearing forests or constructing irrigation structures. Just as with the other input subsidies, they benefit those who need them the least: 'inevitably, subsidized credit schemes in rural areas, even those specifically designed for smallholders, are quickly captured by larger farmers, who are considered to be better risks, more influential, and less costly to serve,' (ibid, p.79). Where credit programmes contribute to the conversion to more capital-intensive forms of agriculture with significant economies of scale, such as ranching, they displace farm labour, thus contributing to the pressures on forestlands, just as is the case with subsidies for mechanization.

Although past experience with agricultural development programmes and subsidies for agricultural inputs in general turned out to be not very encouraging, the need to reform smallholder agriculture into more productive and sustainable land use systems in order to lessen the need for forest-destructive shifting cultivation practices remains. The following section, therefore, after discussing the concept of sustainable development, will focus on agroforestry, a collective name for land use systems widely believed to hold considerable potential as an alternative for current unsustainable forms of agriculture in the tropics.

SUSTAINABLE DEVELOPMENT: THE CASE OF AGROFORESTRY

Since the Brundtland Commission (1987) paid extensive attention to the sustainability aspect of development, the term 'sustainable development' became a common topic in the international policy debate. The Brundtland Commission defined sustainable development as 'meeting the needs of the present generation without compromising the need of future generations.' Thus the sustainable development concept, as compared with previous development concepts, not only takes the need to improve welfare of the present generation into account, but also recognizes that world resources are finite and that wasteful use of existing resources today will cause an unnecessary sacrifice of income and wealth in the future. Sustainable development does not imply that all natural resources should be preserved, however. Successful development will inevitably have an impact on natural resources (World Bank, 1992). Precisely because the need to sacrifice an amount of natural resources is recognized, it has proven difficult to define sustainability more precisely.

In their theoretical approach to the problem Pearce et al. (1990) define sustainability as: 'the general requirement that a vector of development characteristics be non-decreasing over time, where the elements to be included in the vector are open to ethical debate and where the relevant time horizon for practical decision making is similarly indeterminate outside of agreement on intergenerational objectives.'

They subsequently introduce the concept of a constant capital stock defined as the 'requirement that the total value of all capital stocks be held constant, man-made and natural. The basic idea is that future generations would inherit a combined capital stock no smaller than the one in the previous generation.' A similar definition is given by Repetto (1989):

> Sustainable development is a strategy that manages all assets, natural resources, and human resources, as well as financial and physical assets, for increasing long-term wealth and well-being. Sustainable development as a goal rejects policies and practices that support current living standards by depleting the productive base, including natural resources, and that leaves future generations with poorer prospects and greater risks than our own.

This approach leaves societies the choice to focus more on the accumulation of human capital (through education and technological advance) or of man-made physical capital, possibly in exchange for the loss of the value of their natural resources. What matters in the end is 'that the overall productivity of the accumulated capital – including its impact on human health and aesthetic pleasure, as well as on incomes – more than compensates for any loss from depletion of natural capital' (World Bank, 1992). Not everybody would agree with the former definition: some people defend the strong separability between natural and other forms of capital.

A so-called 'constant natural capital stock approach', is advocated by those who argue that aggregate natural capital should be preserved, with losses in one area replenished elsewhere. This approach focuses on and requires estimates of the (economic) value of environmental resources, something which is not always easy if there is an objective assessment at all.

Sustainable agricultural development

Keeping the above general remarks in mind, we now turn to the sustainability issue in a more concrete way, i.e. relative to agricultural development. In order to achieve a more sustainable form of agricultural development in the humid tropics it is generally necessary:

1) that productivity and yields are improved so that pressures on natural forests are relieved;
2) that farming systems in areas already under cultivation are such that the production potential is preserved; and
3) that the value of the land area *under* forests is increased (see also Chapter 3).

Some promising results for the possibility of improving and maintaining productivity on already deforested lands, in order to reduce pressures for additional forest conversion, were obtained in research projects in Brazil, Indonesia and Peru. Based on experiments carried out in Yurimaguas, for example, it was estimated that for every additional hectare with sustainable and high level productivity, an estimated 5-10 hectares a year of tropical rainforests could be saved from the axe of the shifting cultivator (World Bank, 1992). The transition in these experiments from shifting cultivation to sustainable high productivity continuous cultivation is achieved in a number of phases. In the initial phase secondary forest fallows left by slash-and-burn agriculture are taken, and low-input methods are applied. The profit from this initial phase averages $1100 per hectare a year, or a 120 per cent return over total costs (largely labour) for small farmers. With the income generated after several years of the low-input system, a transition can be made to more productive systems, including intensive continuous cropping, legume-based pasture or agroforestry.

Sustainability can be improved through the spread of practices such as conservation tillage, integrated pest management, and the conversion to other varieties. With the conversion to pest-resistant varieties, for example, crop losses in developing countries have already been substantially reduced. Regarding the conditions under which farmers are willing to adapt these technologies and methods the World Bank (1992) states that 'none is more important than the allocation and protection of property rights. In addition, as technologies become more sophisticated, farmer education and strengthening of extension systems are essential'. Integrated pest management, a pest-control system where chemical pesticides are used less often and in smaller amounts, for example, calls for carefully timed, selective spraying of pesticides, backed up by the encouragement of natural predators and more use of resistant varieties and crop rotation. To work, the technique requires on-site research and testing, adaption to particular pests, and sensitivity to socio-economic conditions. Farmers need to be well trained and to receive plenty of expert support, according to the same source.

Although it is not possible to estimate the costs of making agriculture sustainable, some estimates are available of the costs of preventing soil erosion and degradation, and of rehabilitation of degraded areas. Where the capital costs of prevention vary from $50-500 per hectare (sometimes less), rehabilitation may cost from $500 to several thousands of dollars per hectare, depending on the severity of the problem (World Bank, 1992). The costs per hectare of preventive measures on undegraded lands vary with the farming system, the methods used, and the topography; relatively simple ones such as farm forestry and contouring with vetiver grass or other vegetative barriers vary from $50-150, whereas for measures such as terracing, land levelling, earth banks, and the like $200-500 may be required. Since the costs of preventing are comparably small, it is obvious that preference should be given to prevention measures.

The effects on farm output are such that payback periods of investments for prevention can be short (five to ten years or less), on the condition that the participation level of these programmes is high. Public expenditure, therefore, will be needed for research, extension, training, education (including the costs of encouraging community participation in these programmes), and support for infrastructure and afforestation. The World Bank (1992) provides some indicative estimates of the costs of required additional investments, and the benefits of environmental programmes by 2000 for all developing countries. In Table 5.8 those costs and benefits of programmes, contributing either directly or indirectly to a slowdown of deforestation rates, are presented. That is, programmes directed at improved agricultural practices, afforestation and improved forest management, directly contributing to forest conservation, and programmes contributing to a decrease of population growth rates, family planning and female education, thus relieving pressures on forests.

The table shows that the costs of these programmes are relatively small in terms of GDP percentages, especially if the environmental damage of not implementing them is considered. As for improved soil management practices, for example, investments of $10-15 billion a year (0.2-0.3 per cent of GDP) in the 1990s, are required to extend the coverage of improved soil management practices by up to 100 million hectares each year (currently, 1.1 billion hectares are under crops in developing countries, and 2.5 billion hectares are under permanent pasture (World Bank, 1992)). Investment costs may rise by a further $2-3 billion a year if the need for reforestation projects in watershed areas is included (unit costs vary between $500 and $1500 a hectare). Current R&D expenditures on agricultural research need to be expanded by 30-50 per cent in relation to projected levels. In addition, a commensurate increase in finance is required for training and for disseminating the findings of R&D. For family planning and female education programmes it also holds that with a relatively small effort (0.1 and 0.05 per cent of 2000 GDP respectively), the results can be considerable.

The World Bank puts special emphasis on the need to expand R&D expenditures for agricultural research and extension services. Additional efforts for the development of new and higher-yielding cultivars of plants and improved farming systems will be needed because even if existing knowledge is fully exploited, the availability and quality of land and irrigation water will be insufficient to meet demand.

A general definition of a farming system capable of increasing food production for an increasing population on a sustainable basis is given by Pingali (1990): 'a system of farming that closely mimics the dense

Table 5.8 Estimated costs and long-term benefits of selected environmental programmes in developing countries

Programme	*Additional investment in 2000*			*Long-term benefits*
	$ billions p.a.	*% of GDP in 2000[a]*	*% of GDP growth 1990–2000[a]*	
Soil conservation and afforestation, incl. extension and training	15.0–20.0	0.3–0.4	0.7–1.0	Improvements in yields and productivity of agriculture and forests, which increase the economic returns to investment
Additional resources for agricultural and forestry research, in relation to projected levels, and for resource surveys	5.0	0.1	0.2	Lower pressures on natural forests All areas eventually brought under sustainable forms of cultivation and pasture
Family planning (incremental costs of an expanded programme)[b]	7.0	0.1	0.3	Long-term world population stabilizes at 10 billion instead of 12.5 billion
Increasing primary and secondary education for girls[b]	2.5	0.05	0.1	Primary education for girls extended to 15 million more girls, and secondary education to 21 million more Discrimination in education substantially reduced

Source: World Bank 1992, Table 9.1, p.174.

Notes:

a) The GDP of developing countries in 1990 was $3.4 trillion, and it is projected to rise to $5.4 trillion by 2000 (in 1990 prices). The projected GDP growth rate is 4.7 per cent a year;

b) Recurrent expenditures on these items are counted as investments in human resources.

natural vegetation of the humid tropics is what will work in the long run.' Agroforestry systems, whose spread is advocated as a solution to the deforestation problem, fit this description. A recently much promoted method for agricultural development is LEISA (Low External Input Sustainable Agriculture). LEISA, by seeking to optimize the use of local resources, is considered by some as an option feasible for a large number of farmers, because most farmers have limited access to artificial inputs. Agroforestry examples are, among others, explicitly mentioned as promising LEISA techniques for the humid tropics; see for instance Reijntjes et al. (1992). The advocates of LEISA argue that under LEISA conditions, where diverse products are needed and where perennial biomass and functional diversity are of key importance for protecting and reproducing the farm system, agro-ecosystems would ideally approach the climax ecosystem for the site. In the tropics, this would normally be some type of agroforestry system.

For a discussion of other alternative forest management systems, such as collaborative forest management, community forestry and joint forest management, the reader is referred to chapter 3.

Sustainable agriculture: agroforestry

The focus on agroforestry as one of the possible solutions to the deforestation problem is relatively new. As a land use system, agroforestry has long been known and practised, amongst others by farmers in tropical rainforests. One of the ways to stop the deforestation process is to find and establish land use systems that are sustainable. Today different agroforestry practices are considered as ways of attaining sustainable land use, especially in the tropics. 'If integrated into farming systems, trees can help to ensure environmental stability by reducing the pressure on natural forests, maintaining the soil and water resources, mitigating the effects of climatic irregularities and giving products valuable to the social economy.' (Alriksson and Ohlsson, 1990, p.1.)

A question that immediately arises is why, when the condition of sustainability is met and the system has been known for a long time, agroforestry has not been implemented earlier and has not become a widely practised land-use system in the tropics. An important statement in a World Bank paper on the prospects for agroforestry in the tropics strengthens this question: 'A significant feature that emerges from this type of ecological and geographical analysis of tropical agroforestry systems is that, irrespective of the socio-cultural differences in different geographical regions, the major types of agroforestry systems are structurally similar in areas with similar or identical ecological conditions.' (Nair, 1990, p.8.)

Agroforestry is a scientific term for land-use systems that existed long before this term was initiated. Although traditional shifting cultivation can be regarded as agroforestry, it is generally not interpreted as such, see for instance Raintree and Warner (1986, p.39), who describe agroforestry as a systematic approach to the recombination of the basic elements of shifting cultivation into more intensive, sustainable and politically viable forms of land-use.

In the existing literature on agroforestry, definitions of agroforestry, as well as land-use systems which should be regarded as such, vary greatly. The problem with earlier definitions is, as Raintree (1991) comments, that they are rather normative; they state not merely what agroforestry is, but also what it should be: a land management system that is by definition productive, sustainable and culturally appropriate. Lundgren and Raintree (1983) have defined agroforestry in a more neutral way as a collective name for land-use systems and technologies where woody perennials (trees, shrubs, palms, bamboos, etc.) are deliberately used on the same land management unit as agricultural crops and/or animals.

One difficulty with the analyses of agroforestry is that the term 'agroforestry' is applied to farming systems ranging from 'pure' agriculture to 'pure' forestry (Nair, 1990). This is what Vergara (1982) called 'the agriculture-forestry continuum.' Although theoretically the whole range can be defined as agroforestry, in practice there is a grey area between agroforestry and traditional forestry,[2] even nowadays. Moreover, each author has his own way of interpreting the definitive 'range' of agroforestry: Raintree (1991, p.2) argued the use of the term agroforestry in the broadest possible sense and sometimes interchangeably with community forestry, in order to avoid artificial boundaries and to ensure a flexible approach to the integration of trees into rural development efforts.

Existing agroforestry systems can be classified according to their component structure. Woody perennials (trees), are per definition present in agroforestry systems but these systems include at least one other component. Those additional components can be either herbaceous plants (crops) and/or animals. Therefore, there are three possibilities to classify agroforestry systems according to their component structure:

1) agrisilvicultural (crops and trees);
2) silvopastoral (pasture/animals and trees); and
3) agrosilvopastoral (crops, pasture/animals and trees).

The structural classification here is basically an agrisilvicultural one. Furthermore we are specifically interested in agroforestry systems that are appropriate for lowland humid tropics, because they include areas under natural rainforest.

Among the agroforestry systems in use in the humid tropics, shifting cultivation, home gardens, plantation crop combinations, farm woodlots, alley cropping/hedgerow intercropping and Taungya, are the most common. Although the particular systems employed and their spatial distribution are known, data on the relative magnitude of agroforestry as a land-use system are scarce.

Shifting cultivation
As a system of land use which entails the deliberate association of trees with herbaceous field crops in time, traditional shifting cultivation is one of the most ancient, widespread and, until recently, ecologically stable forms of agroforestry (see the beginning of this chapter).

Homegardens
The homegarden is a system of agricultural production conducted largely by household members at or near the residence. Produce from plants and animals in each garden supplements the homegardeners' meals. Surplus may be exchanged, sold or shared beyond the household (Brownrigg, 1985 in Dorland et al., 1988, p.105). Homegardens is a well-known example of agroforestry in all ecological regions in the tropics and subtropics, especially in humid lowlands with high population density. This integral cropping system of agroforestry

---

[2] For example, one FAO paper – Forestry for local community development (1978) – is about the reintroduction of forestry, but the issues that are dealt with are the same as in agroforestry literature.

evolved in places like Java where homegardens contain permanent combinations of annual and perennial food crops and forest trees that form plant communities of several crown levels.[3] There is an important role for the multipurpose tree for economic as well as environmental reasons: it provides animal fodder, food for people, fuel and building materials, and at the same time it stabilizes the soil. Java hill farmers have used this method for hundreds of years while retaining the land's productivity (Vergara, 1982, p.19).

Plantation crop combinations
Tropical perennial plantation crops occupy about 8 per cent of the total arable land in developing countries (Nair, 1990, p.41). Some of those crops produce economic products with a high value – including oil-palm, rubber, coconut, cacao, coffee, tea, cashew and black pepper – for the international market, and are therefore very important, economically and socially, to the countries that produce them. Compared to other agroforestry practices like homegardens, there exists a lot of scientific information about plantations of crops such as rubber, coffee and tea. Contrary to popular belief, a large and growing proportion of perennial-crop cultivation is in the hands of smallholders (Ruthenberg, 1980, p.257).

Farm woodlot
This is a group or block of trees grown mainly to provide construction wood or fuelwood. Crops are generally not grown within the woodlots, but grown adjacent to them.

Alley cropping/hedgerow intercropping
Leguminous trees are grown in rows in cropland with regular spacing between the rows. The main purpose is the use of the leaves for green manure or fodder. Fuelwood is occasionally harvested as a by-product.

Taungya
A little more than a century ago, the Taungya system was developed in Myanmar and spread eventually throughout tropical Asia, Latin America and Africa. In the Taungya system farmers are given temporary access to land and sometimes a modest wage in return for their labour in the planting and care of commercial forest seedlings. The farmers are allowed to integrate farming in forest production although they themselves do not hold tenure rights. Thus, in this system forest plantations and food crops are grown on the same area, but because the forest canopy closes after about three years, food cropping (and farmers) must move to other open areas, where the process starts again. The benefits of the system are that the establishment costs of tree plantations are reduced and that rural dwellers – at least

---

[3] Homegardens are multi-storeyed systems: an association of tall perennials with shorter biannual and annual crops in which the canopies of the crops have a multi-storey structure, allowing an efficient use of sunlight (Steiner, 1982, in Dorland et al., 1988, p.104).

temporarily – are provided with cropland. A considerable disadvantage is that relatively few shifting cultivators can be absorbed. This method is thus well-adapted to forest areas with low population pressure, but it is unsuitable for more densely populated areas that are inhabited by shifting cultivators.

Benefits from agroforestry can broadly be divided into environmental benefits and socio-economic benefits. Environmental benefits from agroforestry are mainly generated by the soil conserving practices of these systems, contributing to a sustainable form of land use. Sustainability of the production system is determined by several criteria, however. Another important criterion for sustainability, and closely related to soil productivity, is erosion control. It is now widely believed that agroforestry holds considerable potential as a major land-management alternative for conserving the soil and maintaining soil fertility and productivity in the tropics. This belief is based on the assumption that trees and other vegetation improve the soil beneath them (Nair, 1990, p.9). There can also be some negative factors as a result of the combined cropping of trees and food crops, for example possible competition trees and food crops for space, sunlight, moisture and nutrients, which may reduce food crop yields.

The major types of agroforestry systems are structurally similar in areas with similar or identical ecological conditions. But the type of agroforestry system found in a particular area is determined only to some extent by agro-ecological factors (Nair, 1990, p.8): 'However, several socio-economic factors, such as human population pressure, availability of labour and proximity to markets, also come into play, resulting in considerable variations among systems operating in similar or identical agro-climatic conditions.'

As a general rule, ecological factors determine the major type of agroforestry system in a given area, but the complexity of the system and the intensity with which it is managed increase in direct proportion to the population intensity and land productivity of the area. In the case of shifting cultivation and Taungya systems, for example, there are numerous variants that are specific to certain socio-economic contexts. Thus socio-economic factors only influence the way a system is used – but not which system.

Socio-economic factors, in areas with identical conditions and proven viable and profitable under these conditions, determine perhaps not only how, but, more importantly, also if an agroforestry system is practised. The choice of a particular production system will be influenced by the socio-economic situation faced by the land-user. It is therefore obvious that feasibility studies on agroforestry systems should, apart from an assessment of the biological/environmental factors, also contain an evaluation of the socio-economic conditions under which these systems are practised.

One of the difficulties in assessing the economic advantages of agroforestry systems has been that past research was directed mainly at the development of new or improved land-use practices, or to cope with such issues as fuelwood shortages, than with increasing the knowledge about the economic contributions of such practices.

Table 5.9 Principal benefits and costs of agroforestry

Benefits and opportunities	Costs and constraints
Maintains or increases site productivity through nutrient recycling and soil protection, at low capital and labour costs.	Reduces output of staple food crops where trees compete for use of arable land and/or depresses crop yields through shade, root competition or allelopathic interaction.
Increases the value of output on a given area of land through spatial or intertemporal intercropping of tree and other species.	Incompatibility of trees with agricultural practices such as free grazing, burning, common fields, etc., which makes it difficult to protect trees.
Diversifies the range of outputs from a given area of land, in order to (a) increase self-sufficiency and/or (b) reduce the risk to income from adverse climatic, biological or market impacts on particular crops.	Trees can impede cultivation of monocrops and introduction of mechanization, and so (a) increase labour costs in situations where the latter is appropriate and/or (b) inhibit advances in farming practices.
Spreads the needs for labour inputs more evenly seasonally, so reducing the effects of sharp peaks and troughs in activity characteristic of tropical agriculture.	Where the planting season is very restricted e.g. in arid and semi-arid conditions, demands on available labour for crop establishment may prevent tree planting.
Provides productive applications for underutilized land, labour or capital.	The relatively long production period of trees delays returns beyond what may be tenable for poor farmers, and increases the risk to them of insecurity of tenure.
Creates capital stocks available for intermittent costs or unforeseen contingencies.	

Source: Arnold (1987), p.175.

Agroecological research has clearly pointed out the viability of agroforestry. Until recently, however, economic analyses of agroforestry systems were relatively scarce, but are now becoming more easily available.

In the literature on positive and negative economic features of agroforestry the analyses of Arnold (1987) are often referred to. (See Alriksson and Ohlsson, 1990, and Nair, 1990.) Arnold examined some existing agroforestry practices – mainly homegardens and farm woodlots – in order to try and identify the economic considerations that have caused farmers to adopt them.

Labour requirements

One of the most important socio-economic aspects of agroforestry is the labour absorption capacity of the system. The labour intensity requirements of the agroforestry system is one of the key decisive factors in moving from traditional shifting cultivation practices to intensive agroforestry systems. Agroforestry can provide a productive application for underutilized labour. According to Ruthenberg (1980, p.103), in most degenerated shifting systems only about one-third to two-thirds of the available labour capacity is absorbed in field work. Nevertheless the shifting cultivators, regardless of the degree of underemployment, unanimously feel that labour shortage is the most important factor limiting output. The fact that important labour shortages within family holdings and a high degree of underemployment go hand in hand may be traced back to a number of major factors (Alriksson and Ohlsson, 1990 and Nair, 1990):

- labour requirement peaks, particularly with arable crops, occur seasonally;
- agricultural work is not the only task that has to be performed by smallholdings. General household work seems to be much more time-consuming than field work;
- the labour capacity of the various persons in the household is not utilized equally, because of the traditional concepts of the way in which work is divided between man and women;
- the low efficiency per hour of work – smallholders are accustomed to making a concentrated, sustained effort only in connection with a few procedures, such as felling and clearing, and the other jobs have traditionally been done in a leisurely manner;
- the extent of underemployment is calculated according to norms of working hours per year and per man equivalent – the smallholders who refer to difficulties in the labour economy are typically not familiar with such norms.

Systems that are in a transitional phase between shifting and permanent land use are facing problems in their labour economy: for example if the transition to cash cropping, with its consequent labour demands, takes place within a fairly short period before the farmers are accustomed to longer hours of work per day. Consequently, they are interested in labour-saving innovations, even more than in yield-increasing innovations.

Assuming that farming families have somehow planned their labour input (according to time and task), it is obvious that additional labour requirements for persons already fully occupied at peak labour seasons will be important – and can be a drawback – to the adoption of a new practice. With respect to labour peaks agroforestry systems are often said to have the advantage of helping to spread the use of labour supplied by members of a farming family more evenly throughout the year.

Land use and land-tenure

Many existing agroforestry systems require both high labour intensity and high land-use intensity (Raintree, 1983, in Alrikkson and Ohlson, 1990, p.9). High population growth increases food demands and when land becomes drastically scarce, it is common that production on a short-term basis takes precedence over long-term values with a breakdown of sustainability as a result. Here we are back to the roots of our deforestation problem. If the increase in pressure on land is a slower process, agroforestry systems like home gardens might be a spontaneous response.

In areas where land tenure systems do not guarantee continued ownership and control of land, long-term agricultural strategies like agroforestry will not easily be adopted. It is obvious that the incentive for investing in soil fertility improvement for future use of the land is low unless the benefits accrue to the tree planter. A distinction should be made between rights over land and rights over trees, because those two do not always coincide. Tree tenure issues include the right to own or inherit trees, the right to plant trees, the right to use trees and tree products and the right to dispose of tree products (Fortman, 1988, in Nair, 1990). These various rights differ widely across cultural zones and have major influence on the social acceptability of agroforestry initiatives.

Marketability of tree products

Of crucial importance to the local people is the direct income they can generate from a particular land-use system. The processing and/or sale of agricultural commodities, and the rural industries based on these commodities, are essential sources of off-farm income for many farming societies. Studies of FB-SSI (e.g. FAO, 1987) indicate that these enterprises are important employers of rural people, especially resource-poor and landless people. However, the problem is that such enterprises have poor access to markets and raw materials and inadequate organizational and management skills. This means that if agroforestry is to contribute to rural development, additional policies to strengthen or establish appropriate market infrastructures are necessary as well as the development of the necessary skills.

Accessibility to the capital market

As discussed at the start of this chapter, socio-economic and institutional restraints play an important role in the degeneration of shifting cultivation. In order to move to an evolutionary type of agroforestry investments are needed. But since credit facilities are too poor, especially for small farmers, the low investment problem cannot be solved.

The above general socio-economic conditions determine whether or not a transition to agroforestry systems is likely to occur. The adoption of agroforestry systems by individual farmers in the tropics is determined by their perception of the costs and benefits of growing trees on their farms. Arnold (1987) examined some selected existing agroforestry systems – mainly homegardens and farm woodlots – in order to try and identify the economic considerations that have caused farmers to adopt them. The results are represented in Table 5.10.

With respect to homegardens the overall process in the selected cases appeared to be the same. The farmers' response to increasing population pressure and decreasing land availability is to turn more and more to agroforestry systems. Initially this is because these permit more sustainable intensification of land use and higher returns from available labour inputs than alternative uses. However, as population density continues to increase, yields and return to labour eventually decline to the point where farmers have to turn increasingly to non-farm sources of income. The agroforestry system can be maintained when trees and other perennials requiring only low labour inputs come to form the main component, thus as a low-input, low-management type of land use. This process changes at the point where farmers are able to inject substantial capital into their systems and intensify land use further through purchased inputs of fertilizer, herbicides etc.[4] This reduces the importance of multipurpose trees in soil-nutrient maintenance and weed suppression, causing them to be an impediment rather than a complement to agriculture. Trees are then cultivated only where they are competitive as cash crops. Nevertheless, the subsequent displacement of agroforestry practices seems to confirm that, in the absence of capital, farmers had been employing trees primarily to provide substitutes for purchased inputs, and as crops requiring lower inputs than agricultural crops.

With respect to woodlots Arnold (1987, p.180) concludes that the decision to grow trees has been influenced by two main factors. One is the high cost of labour and capital, and the advantages tree cultivation offers in this respect because of its low input requirements. The other is the prominent part that income generation, as distinct from food production, plays in the farmers' production objectives.

It seems that farmers employ agroforestry practices primarily because they perceive these as being the most efficient way of meeting their production goals using the resources of land, labour and capital available to them. Equally, needs and opportunities linked to off-farm employment are likely to play a role in their assessment of the most efficient farm strategy to choose.

An important constraint to the adoptability is the long production period of most tree species. It takes a long time before the investment of tree planting starts to generate income. Furthermore, trees do not produce staple food. An investment in trees which might not start to bring in money for 15–20 years or even longer, does not always appear as a choice with high priority to a farmer who has scarce land resources, even if it may have a positive impact on sustainability in a long-term perspective (Arnold, 1983).

Under conditions where agroforestry is extended as a new way of production, it is important to evaluate what will determine the adoptability of the new system. Raintree (1983, in Alriksson and Ohlsson, 1990, p.9) states that three main types of factors will be the conclusive arguments in the decision process:

1) the characteristics of the potential adopters (including their situational constraints and potentials);

---

[4] Arnold, 1987, based on a study of tree cropping in the homesteads in Kerala, India, by Nair and Krishnankutty, 1984.

Table 5.10 Economic factors affecting adoption of agroforestry practices in selected situations

Agroforestry systems	Constraints/ opportunities	Farmer response	Contribution of agroforestry
Homegardens, Java	Declining landholding size, minimal or no rice paddy, minimal capital	Increase food and income output from homegardens	Highest returns for land from increasing labor inputs, flexibility of outputs face of changing needs and opportunities
	Further fall in landholding size below level able to meet basic food needs	Transfer labour to off-farm employment	Most productive and stable use of land with reduced labour inputs
Compound farms[1], Nigeria	Declining landholding size and site productivity, minimal capital	Concentrate resources in compound area, raise income-producing component and off-farm employment	Improves productivity, highest returns to labour, flexibility
Homegardens, Kerala	Declining landholding size, minimal capital	Bring fallow into land use, intensify homegarden management	Multipurpose trees maintain site productivity/ contribute to food and income
	Capital inputs substantially increasd	Transfer land use to high value cash crops, substitute fertilizer and herbicide for mulch and shade	Trees removed unless value crop producers
Farm woodlots, Kenya	Farm size below basic needs level, minimal capital, growing labour shortage	Low-input low management pole cash crops, off-farm employment	Lower capital input than alternative crops and higher returns to labour
Farm woodlots, Philippines	Abundant land, limited labour	Put land under pulpwood crop	Expands area under cultivation, increases returns to family labour

[1] also known as homegardens

Source: Arnold, 1987.

2) the manner in which the innovations are communicated to them (i.e. the extension process);
3) the nature of the innovation itself (i.e. attributes of the proposed technology).

Economic evaluation of agroforestry technologies[5]

Despite all the excitement about agroforestry, the important question remains: how realistic is the contribution of this type of land-use to stop or slow down the deforestation process in the humid tropics? In the existing literature on agroforestry, there are two main issues which the authors seem to agree upon. In the first place, there is a wide consensus on the environmental/physical benefits – notably sustainability – of agroforestry. A second, frequently repeated, conclusion is that more and better economic studies in agroforestry are urgently needed. Here lies the problem of answering our question. Moreover, conclusions in the studies available so far about the economic viability of agroforestry systems in tropical forest area, do not always point in the same direction.

Since labour requirement is a key factor in the adoptability of the system, a substantial labour absorbtion capacity of agroforestry technologies would be an important advantage in diminishing the effect of population pressure on the rainforest. Also the issue of land productivity rates is of crucial importance. If with the same amount of labour, less agricultural land is necessary for the same output, this would also ease the pressure to cut the rainforests.

It appears that more labour and yield data have become available over the last five years. Some conclusions from the documents on labour requirement are summarized below:

- A study of resource productivity in Taungya farms in Nigeria concluded that labour is not used efficiently and its input should be reduced by farmers (Abu, 1989, in ICRAF, 1991, p.31);
- In alley cropping compared to the sole crop systems (Ghana), the labour requirements appears to be higher during the early stages, due to extra labour needed for leucaena establishment. In later stages however, the labour required for hedge pruning will be compensated for by less weeding needed in the cropped alleys (Balasubramanian, 1983, in ICRAF, 1991, p.40). Similar conclusions are found in a study on mixed and zonal systems in Malaysia (Hoekstra, 1984, in ICRAF, 1991, p.89);
- In a (one year-period) study conducted to assess the socio-economic advantages of various integral agroforestry systems carried out in Papua New Guinea, it was found that labour employment is more evenly spread over the year in the agroforestry system, and that the seasonal fluctuations in labour and income are reduced by 60 per cent. This is caused by the complementarity of labour use in the annual and perennial components of the agroforestry system. Labour absorption has increased 52 per cent in agroforestry. The returns per farm and per capita also increase substantially in this system. However, the returns per ha or man day of work are not

---

[5] This term corresponds more or less with the definition of agroforestry practices, but it is furthermore explicitly mentioned that an agroforestry technology is designed to perform specific functions through appropriate management inputs. Specific agroforestry technologies can be interpreted as agroforestry systems.

always higher. According to the authors this might be due to a marginal return of these resources which is lower than the average revenue and by using more of these resources profits are higher but returns to these resource lower. Possibly, farmers might also strive for a particular income target only (Flores and Vergara, 1986, in ICRAF, 1991, p.76);

- The farmer needs to be adequately compensated for his labour in order to accept an agroforestry system. An ex post cost-benefit analysis (CBA) of an agroforestry programme in Haiti showed that 15 per cent of the farmers involved in the project had negative returns. The author (Hosier, 1989, in ICRAF, 1991, p.98) attributes this to static assumptions on the cropping pattern leading to a too-high opportunity cost of land, lack of information, or the possibility that agroforestry may not be the best solution for these farmers. In general, important issues in CBA of agroforestry are opportunity cost of land, the discount rate, and the value of labour which is never zero;

- In a study (also used by Arnold, 1987 – see p.130 – and Dorland et al., 1988) of the effects of population pressure on traditional farming systems in eastern Nigeria, Lagemann (1977) found that soil fertility declines with higher population density and that farmers react by concentrating production on small compounds with high input of mulch, manure and household refuse. The size of labour input did not differ significantly between the outer fields and the compounds. The average gross returns to labour from compound farming were four to eight times higher than from the outer fields. The labour productivity is found to diminish with increased population densities;

- An economic evaluation of alley cropping leucaena-maize in Nigeria (Ngambeki, 1982, in ICRAF, 1991, p.133) led to the conclusion that although labour costs increased because of the cutting and pruning of leucaena, the economic benefits of leucaena showed a much higher increase. The author drew similar conclusions after further studies (1985); on the average the hedges increased labour inputs by 52 per cent per season but increased yield by 68 per cent;

- Cost–benefit analyses of the use of three labour alternatives – direct permanent, direct casual and Taungya labour – in pulpwood plantation establishments in the tropical rainforest of Nigeria, showed that cost per hectare is lowest with Taungya labour and highest with permanent labour. Savings in labour costs with Taungya are 30 and 47 per cent over casual and permanent labour respectively. The net present value per hectare of investment is highest when Taungya labour is used and lowest (and negative) with permanent labour (Nwonwu, 1987, in ICRAF, 1991, p.135);

- Contour planting (including grass, fodder species, leguminous creepers and trees) as a means to prevent soil erosion in Tanzania, did provide additional benefits in sustaining crop yields through soil erosion control. But while returns to labour were higher than for subsistence crops, they were lower than for cash crops (Taube, 1988, in ICRAF, 1991, p.165);

- Verinumbe (1984, in ICRAF, 1991) shows that the availability of hired labour determines which alley-cropping species are chosen. The most profitable option is the one where hired labour is available at relatively low cost;
- Raintree (1983) argues that the task for technology developers and extensionists is to match up the labour requirements and land-use intensities of candidate technologies with the corresponding levels of labour availability and land-use intensity most appropriate to the farming system. Therefore, if agroforestry technologies are introduced it would be best to use a phased approach to intensification and develop a succession of technologies;
- Dewees (1989) tries to explain – on the basis of historical changes in farmers' use of trees – the current importance of tree planting in the economic strategies of smallholders in Kenya. The thesis of the study is that the major incentives for tree growing stem from poorly functioning factor markets leading farmers to seek an alternate source of capital and labour-extensive income-earning options.

Some final remarks on the matter of tenure right are in order. Tenure rights are expected to play an important role in the adoption of agroforestry systems. A selection of the conclusions of the literature on this subject is presented here:

- Meinstock and Vergara (1987, in ICRAF, 1991) investigate the distinction between the right to land and the right to plant. Conflicts are discussed between the traditional dichotomy of land and plant rights and government policy. It is concluded that the perceptional separation of land and plant ownership needs to be explored if agroforestry practices are not only to be ecologically and economically feasible but also culturally acceptable;
- Measures should be taken to legalize individual landownership to make agroforestry systems which are profitable on paper and feasible in practice (Openshaw and Morris, 1979, in ICRAF, 1991, p.137);
- A multi dimensional approach to assess the impact of agroforestry in the Philippines was carried out by Rola (1989, in ICRAF, 1991). Data include tenurial status, income, labour use, soil fertility and organic matter of both agroforestry and non-agroforestry soils. A regression analysis integrating these multi-disciplinary aspects was carried out. It shows the positive impact of tenure status on the number of leucaena planted for sale, which in turn relate strongly to higher income, more labour use and less erosion. Other relations found are not causal and basically derive from these.

How is population pressure related to agroforestry?

Not many documents explicitly provide conclusions on the link between population pressure and agroforestry. In this respect conclusions about labour requirements are important. Nevertheless some can be mentioned:

- As population pressure increases, the intensity of the multiple cropping of food crops and trees first increases but then decreases again in favour of staple food crops after a certain threshold value has been reached (Java: Berenschot et al., 1986, in ICRAF, 1991, p.43).
- O'kting'ati and Mongi (1986; in ICRAF, 1991), studying agroforestry and the small farmer (Tanzania), concluded that agroforestry practices (trees mixed in annual and perennial cropland) have helped to maintain a large and expanding population in an area very prone to erosion and soil degradation (slopes of Mt. Kilimanjaro). Unfortunately we have no exact details about how this was achieved.

Is agroforestry profitable?

Conclusions about the general profitability of agroforestry systems in the tropics show a similar diversified picture as with labour requirements:

- *profitable*
  - Ecuador: the inclusion of marketable trees in coffee plots (the main crop) appears to be an attractive alternative because of their low labour requirements and potential value (Estrada et al., 1988, in ICRAF, 1991, p.67).
  - According to Hoekstra (1987, in ICRAF, 1991) the potential in Africa for agroforestry or other tree-based land use systems is good because there is a great demand for fuelwood, small timber, and tree fodder.
- *profitable but*
  - Intercropping/fallow systems would be very profitable to farmers according to a study by Akachuka (1985), which provides a detailed breakdown of costs, benefits and management over time. However, results are difficult to interpret due to the exclusion of non-cash costs, opportunity costs and other factors.
- *not profitable*
  - de Graaff and Dedwiwarsito (1987) on the Konto River Project in Indonesia; one of the activities consists of Taungya reforestation schemes in which mostly landless households are given a woodlot of 0.25 ha, which they initially intercrop. Direct on-farm benefits will not offset the high cost of terracing to farmers; soils are deep so erosion is not noticeable to farmers. An attempt is made to assess the downstream measures.
  - A comparison between the cash flow/benefit-cost ratios for an agroforestry systems (improved tree fallow) and pure crop and pure tree models, was done for smallholders in Nigeria (Osemeobo, 1989, in ICRAF, 1991, p.138). Although the author concludes differently and in spite of the low discount rate used, the data show that for a small farmer this agroforestry system is not any better than the pure crop system.

Agroforestry research so far has focused mainly on Africa. A survey of project experience of agroforestry in Africa (Kerkhof, 1990) – including 21 projects – was carried out to find out what has worked, and what has not worked, under practical conditions. One of the conclusions was: 'On the key question of sustainability, few projects have yet reached the stage where they can confidently predict that the changes they have introduced will continue, and spread, after the project has finished.' (Kerkhof, 1990, p.10). Here, the issue of *replicability* should be introduced, which is an important criterion to measure the success of a project. The problems being addressed by agroforestry are widespread. They will not be solved by projects which cover a few tens or even thousands of hectares. If agroforestry is to have an impact, the role of projects must be to develop techniques and approaches which are spontaneously replicated by local people on a wider scale.

Although lessons from experience in Africa (and Asia) are useful in examining the potential contribution of agroforestry to stop the deforestation process of tropical rainforests, socio-economic and institutional differences in tropical Latin America call for evaluation of projects experience in this region. These are becoming more and more available, as for instance in an economic analysis of improved agroforestry practices in the Amazon lowlands of Ecuador in Ramirez et al. 1992, pp.65–86. In this study a pilot project of Ecuador's Ministry of Agriculture, to promote improved agroforestry practices in already cleared lands of Ecuador's Amazon lowlands in order to encourage regeneration of deforested areas, is analysed.

According to the authors, improved agroforestry practices in the Amazon lowlands of Ecuador are not only technically feasible but also economically viable and expected to be widely adopted by colonists in the future. However, this expectation can only be partly based on the results of the study, because the adoption rates of newly introduced crops were not very impressive. Furthermore, they state that these practices help increase labour productivity and save on labour and external inputs to control weeds. Therefore, new practices are shown to be more profitable and socially desirable to colonists than traditional ones, by requiring very small investments at the margin, improving cash income distribution over time, and increasing returns to labour (ibid., p.84).

However, it is also remarked that improved agroforestry systems appear quite attractive and could therefore lead to even greater population pressure on the primary forest, if there is no continuity with this effort for planning and resource management. This may be expected since forest land is a 'free-access' resource. That this situation is not hypothetical is shown by present legislation which requires colonists to clear primary forest to establish extensive forms of agriculture before claiming land titles.

From this case-study, as from others, the positive potential of agroforestry comes to the fore. The potential of these land-use systems as a way of reducing the pressures to convert natural forests for agricultural purposes accrues from their capacity of maintaining soil resources as well as from labour absorption and land productivity capacities. Although promising, it is also clear that socio-economic and institutional (tenure) constraints play a major role in the adoption of agroforestry practices in land-use systems in the tropics. Because of these constraints it is highly unlikely that without major efforts in the field of land reform and rural development, a large-scale spontaneous transition to agroforestry by small-scale farmers will occur.

6

SOCIO-ECONOMIC CHARACTERISTICS OF THE MAIN TROPICAL REGIONS

BASIC ECONOMIC INDICATORS

Deforestation has in many countries provided a temporary escape valve from economic and social hardship. Very often deforestation only offers a respite from development pressures, pressures that can be dealt with effectively only at a more fundamental level. If such a fundamental approach is to be taken seriously, this in any case begs for a more complete analysis whereby the various economic linkages are specified.

Indeed, rapid deforestation in the tropics during the 1980s has generally been linked to the exceptionally difficult economic conditions most tropical countries face. Indonesia's drive to export timber products can for example be interpreted as a conscious effort to offset its lower petroleum earnings and protect its development programme from further cutbacks. Many of the most heavily debt-burdened countries are coincidentally those with most of the remaining tropical forests.

So, deforestation cannot be separated from the overall economic climate in the tropical regions, in the sense that for instance a more favourable economic climate could reduce the pressures on the remaining tropical forests of unemployment, poverty and population growth. In the 1980s for example, economic growth in most of the tropical regions failed to outpace the increase in the labour force. Employment in the organized urban sector often stagnated and declined; and real wages plummeted in the informal urban labour market. Instead of the usual rural-to-urban migration, there was a pile-up in the agricultural sector.

In Brazil, for example, the agricultural labour force grew by four per cent a year between 1981 and 1984, compared with a growth rate of only 0.6 per cent between 1971 and 1976; agricultural wages fell almost 40 per cent in real terms between 1981 and 1985. With no alternative, given the concentration of agricultural land in large holdings and the absence of jobs, rural households migrated to the frontier in increasing numbers.

(Repetto, 1990, p.23)

So, land use in the tropical areas is strongly linked with such factors as population patterns, migration within and between the countries

137

involved, and their overall economic performance. In this chapter some general information will be given on the main economic variables underlying the tropical deforestation processes. This will provide an economic background against which tropical deforestation takes place.

> Poverty is one of the greatest threats to the environment. In poor countries, poverty often causes deforestation . . .. And this environmental damage reinforces poverty. Many choices that degrade the environment are made in the developing countries because of the imperative of immediate survival, not because of a lack of concern for the future. Any plans of action for environmental improvement must therefore include programmes to reduce poverty in the developing world.
>
> (UNDP, 1990, p.7)

Tables 6.1a–c provide some general information on population, GNP and external indebtedness of the forty main tropical countries. What are the main broad development characteristics of these countries, and to what extent should a distinction be made between the various main regions, tropical Latin America, tropical Africa and tropical Asia?

A first conclusion from Table 6.1. is that the total population in 1992 living in the tropical regions was roughly 1 billion, or some one-fifth of the world population: some 44 per cent of which lived in Asia, 32 per cent in Africa and the remaining 24 per cent in Latin America. The population pressure in these regions is considerable, however. It is expected that by the year 2000 the total population in the tropical area will already have increased to more than 1200 million people. By then the share of Asia and Latin America will have declined to 43 and 23 per cent respectively, and that of Africa will have increased to 34 per cent. This trend is expected to continue with a doubling of the population in these areas in 2025 to almost 2 billion people, with a further increase of the share of Africa to over 43 per cent, and a further decline of the Asian and Latin American shares to 35 and 21 per cent, respectively.

In addition to the expected doubling of the population in the tropical areas in the 1990–2025 period, urbanization is expected to further increase from the presently already fairly high figures, especially in Latin America (75 per cent). This can put additional unforeseen pressure on the sustainability targets in these regions and, albeit indirectly, aggravate deforestation and other environmental decline. If by 2000 the urban population will have increased in tropical Latin America from the current 187m to some 223m, and the corresponding increase in tropical Africa is from 105m now to more than 170m in 2000, and in tropical Asia from some 134m to over 192m, one may wonder if one can ever prevent these processes from putting further pressure on land use and deforestation.

Another characteristic of these tropical countries is their rather low GNP per capita, in particular in Africa (with the exception of Gabon)[1]. In most tropical countries in Africa GNP per capita in 1992 was less

---

[1] This general impression does not change when a Human Development Indicator, e.g. as it has been designed by UNDP, is employed as the welfare measure rather than the GNP per capita: in the major part of the countries considered this index is less than 0.5.

Table 6.1 Basic economic indicators

6.1a Tropical Latin America

Country	Population			% urban		GDP 1992 ($m)	GNP per capita		External debt		Human development index[1] 1992
	Million		Average annual growth				$ 1992	% growth 1980–92	$m 1992	Debt/GDP 1992	
	1992	2000	1992–2000	1992	2000						
Total/average	249.2	282.4	1.6	75	79		2,282				
Bolivia	8.0	9.0	2.4	52	58	5270	680	-1.5	4243	0.805	0.530
Brazil	154.0	172.0	1.4	77	81	360,405	2770	0.4	121,110	0.336	0.756
Colombia	33.0	37.0	1.4	71	75	48,583	1330	1.4	17,204	0.354	0.813
Ecuador	11.0	13.0	2.0	58	64	12,681	1070	-0.3	12,280	0.968	0.718
Guyana	0.8	0.9	1.1	34	43	n.a.	330	-5.6	n.a.	n.a.	0.580
Peru	22.0	26.0	1.8	71	75	22,100	950	-2.8	20,293	0.918	0.642
Surinam	0.4	0.5	1.7	43	54	n.a.	4280	-3.6	n.a.	n.a.	0.677
Venezuela	20.0	24.0	2.2	91	94	61,137	2910	-0.8	37,193	0.608	0.820

Sources: World Bank (1994), Tables 1, 3, 20 and 25; UNDP (1994), Tables 1, 20, 22, 23 and 26.

Note: 1) Human Development Index = index with a desirable level of 1, measuring the development level by three indicators: life expectancy; literacy; and the purchasing power to fulfil basic needs. The desirable level of 1 is almost reached in a number of developed countries as, for example, Japan and the Netherlands.

6.1b Tropical Africa

Country	Population Million 1992	Population Million 2000	Population Average annual growth 1992–2000	% urban 1992	% urban 2000	GDP 1992 ($m)	GNP per capita $ 1992	GNP per capita % growth 1980–92	External debt $m 1992	External debt Debt/GDP 1992	Human development index[1] 1992
Total/average	329.7	420.8	3.1	32	41	n.a.	n.a.	n.a.	n.a	n.a.	
Angola	9.9	13.1	3.5	27	36		n.a.				0.271
Benin	4.0	6.3	3.0	40	45	2181	410	−0.7	1367	0.627	0.261
Cameroon	12.2	15.3	2.8	42	51	10,397	820	−1.5	6554	0.630	0.447
Central Africa	3.2	3.9	2.5	48	56	1251	410	−1.5	901	0.720	0.249
Congo	2.4	3.0	2.9	42	47	2816	1030	−0.8	4751	1.687	0.461
Cote d'Ivoire	12.9	17.1	3.5	42	47	8726	670	−4.7	17,997	2.062	0.370
Eq. Guinea	0.4	0.5	2.5	29	37	n.a.	330	n.a.	n.a.	n.a.	0.276
Gabon	1.2	1.6	3.2	47	54	5913	4450	−3.7	3798	0.642	0.525
Ghana	16.0	20.2	2.9	35	38	6884	450	−0.1	4275	0.621	0.382
Guinea	6.1	7.8	3.0	27	33	3233	510	n.a.	2651	0.820	0.191
Guinea-Buissau	1.0	1.2	2.1	20	25	220	220	1.6	634	2.882	0.224
Kenya	25.3	32.8	3.3	25	32	6884	310	0.2	6367	0.925	0.434
Liberia	2.8	3.6	3.2	47	58	n.a.	≤675	n.a.	n.a.	n.a.	0.317
Madagascar	12.9	16.6	3.2	25	31	2767	230	−2.4	4385	1.585	0.396
Nigeria	115.9	147.7	5.1	37	43	29,667	320	−0.4	30,959	1.044	0.348
Senegal	7.8	9.6	2.7	41	45	6277	780	0.1	3607	0.575	0.322
Sierra Leone	4.4	5.4	2.6	31	40	634	160	−1.4	1265	1.995	0.209
Togo	3.8	4.8	3.1	29	33	1611	390	−1.8	1356	0.842	0.311
Uganda	18.7	23.4	2.8	12	14	2998	170	n.a.	2997	1.000	0.272
Tanzania	27.9	35.9	3.2	22	47	2345	110	0.0	6715	2.864	0.306
Zaire	40	51	3.1	29	46	n.a.	≤675	−1.8	n.a.	n.a.	0.341

Sources: World Bank (1994), Tables 1, 3, 20 and 25; UNDP (1994), Tables 1, 20, 22, 23 and 26.

Note: 1) Human Development Index = index with a desirable level of 1, measuring the development level by three indicators: life expectancy; literacy; and the purchasing power to fulfil basic needs. The desirable level of 1 is almost reached in a number of developed countries as, for example, Japan and the Netherlands.

6.1c Tropical Asia

Country	Population			% urban		GDP 1992 ($m)	GNP per capita		External debt		Human development index[1] 1992
	Million		Average annual growth						$m	Debt/GDP	
	1992	2000	1992–2000	1992	2000	$ 1992	$ 1992	% growth 1980–92	1992	1992	1992
Total/average	454.1	522.4	1.8	30	37						
Indonesia	191.2	218	1.7	30	40	126,364	670	4.0	84,385	0.668	0.586
Malaysia	18.8	22.3	2.1	45	51	57,568	2790	3.2	19,837	0.345	0.794
Philippines	65.2	76.1	2.0	44	49	52,462	770	-1.0	32,498	0.619	0.621
Thailand	56.1	61.2	1.1	23	29	110,337	1840	6.0	39,424	0.357	0.798
Lao Dem. Rep.	4.5	5.6	2.8	20	25	1195	250	n.a.	1952	1.633	0.385
Vietnam	69.5	81.5	2.0	20	27	n.a.	≤675	n.a.	n.a.	n.a.	0.514
Myanmar	43.7	51.6	2.1	25	28	37,749	n.a.	n.a.	5326	0.141	0.406
Fiji	0.7	0.8	1.0	37	43	n.a.	2010	0.3	n.a.	n.a.	0.787
Papua New. Guinea	4.1	4.9	2.3	16	20	4228	950	0.0	3265	0.772	0.408
Solomon Isl.	0.3	0.4	3.3	8	n.a.	n.a.	710	3.3	n.a.	n.a.	0.434

Sources: World Bank (1994), Tables 1, 3, 20 and 25; UNDP (1994), Tables 1, 20, 22, 23 and 26.

Note: 1) Human Development Index = index with a desirable level of 1, measuring the development level by three indicators: life expectancy; literacy; and the purchasing power to fulfil basic needs. The desirable level of 1 is almost reached in a number of developed countries as, for example, Japan and the Netherlands.

than $500; in Asia there was quite some variety with a per capita GNP varying between $250 in Lao Democratic Republic to over $2000 in Malaysia and Fiji; per capita GNP in tropical Latin America generally is well over $1000, with the exception of Guyana and Bolivia. The major part of all tropical countries considered was confronted with a declining GNP per capita during the period considered, 1980–1992. The only clear exceptions were Brazil and Colombia in Latin America, Guinea, Kenya and Senegal in Africa, and all Asian tropical countries with available data, except for the Philippines.

The general picture that holds for many of the tropical regions is one of not only low levels of per capita income but also of levels that often have declined during the past decade. These developments cannot be separated from the deforestation processes described in Part I.

A final characteristic of the tropical countries is their fairly high external indebtedness. If their external debt is expressed as a percentage of GDP, one seldomly finds shares less than 0.5. It should be recognized that much depends on the debt terms, e.g. whether the debts are private or public, etc. However, it is clear that almost all of the countries have become highly dependent on foreign financial inflows, and that debt servicing obligations aggravate their external constraints. If, for instance, the total debt/GDP ratio is 0.5 and the real interest rate 2 per cent per annum – a rather low estimate – 1 per cent of GDP will continuously be required for external transfers due to debt servicing. It is hard to imagine that the generally weak external position of the tropical countries will not in one way or another affect deforestation. Either the need to earn foreign exchange can promote rapid timber extraction and exports, or the foreign exchange bottlenecks can induce the country to maintain self sufficiency in food production, and expand the area for agricultural use accordingly.

POPULATION GROWTH IN DEVELOPING COUNTRIES

Population patterns, although crucial to deforestation processes, should not be viewed in isolation. A decomposition of environmental impact into its underlying factors can be expressed by the following identity (Harrison, 1991):

I = P.A.T.

with:

I = environmental impact
P = population
A = affluence or consumption
T = technology

The identity makes clear that as a general principle three main factors can be held responsible for environmental impact (given the regeneration capacity of nature, the institutional setting, and the relevant policy measures): population, and per capita consumption or affluence, each usually positively related to the use of resources and hence of the environment, and technology that can either aggravate the environmental problems, or instead contribute in solving natural resource and environmental restraints.

The merit of the above straightforward decomposition identity is that it puts the contribution of population development to environmental damage into its proper perspective. It makes clear that an increasing population pressure and/or an increasing level of per capita consumption will only have adverse environmental impact if technology development is unable to offset the environmental impact of the first two factors, insofar as the burden on the environment surpasses natural regeneration rates.

The identity can also be used to make the opposite point, however, i.e. that an adverse environmental impact can, under certain circumstances, also take place without rapid population growth, for instance if affluence increases too rapidly and/or technology change critically affects the environment in a non-positive way. For if the opposite were true – namely that environmental decline is incompatible with low population pressure – very little environmental degradation would be expected in the low-population-growth industrialized countries. This is obviously not the case. The same equally applies in developing countries: even if they are sparsely populated or have a relatively low population increase, environmental damage nevertheless can be serious, e.g. due to internal factors such as the inequality in access to natural resources, particularly agricultural land; or the breakdown of traditional resource management systems under external pressures.

The above basic approach is also reflected in the discussions on the sustainability concept and on the concept of 'population supporting capacity' or 'carrying capacity.' With respect to this concept one again can discern a variety of views, similar to the different points of view held with respect to the definition of forest resources and deforestation, discussed in Chapter 1. Some observers, many of them ecologists, view the carrying capacity as a key constraint to population growth. Others, mostly economists, consider carrying capacity such a flexible concept – subject to virtually unlimited expansion through technology and policy interventions – that it sometimes ceases to have much operational value at all. These different points of view are expressed in the variety of definitions of the carrying capacity concept. A typical example of a environmentalist/biologist definition is the following: 'The carrying capacity of a particular region is the maximum population of a given species that can be supported indefinitely, allowing for seasonal and random changes, without any degradation of the natural resource base that would diminish this maximum population in the future.' (Mahar, 1985, p.45).

According to this definition, that seems to be designed for non-human species in particular, the carrying capacity should be viewed as a concept linked with a population of a certain species the size of which can be determined objectively; maintaining the carrying capacity is subsequently associated with the ability to maintain this population size. This definition can be equally as applicable to men as it is to non-human species.

Estimates have been made in the literature, more or less in line with this definition, to assess the carrying capacity at the national level of developing countries. A recent example is Hinrichsen (1991) who elaborated on the carrying capacity of the agricultural base of Pakistan, the Philippines, and Mexico, and referred to a FAO/UNFPA study on Kenya. In the latter study it was for instance estimated that the population supporting capacity of Kenya's farm- and rangelands cannot exceed 51m people, even with high level farming inputs (projected population by 2025: 80m).

A completely different definition of carrying capacity is given by Myers, who applies the concept specifically to people. He defines the planet's population supporting capacity more in line with the economists' point of view as the 'number of people that the planet can support without irreversibly reducing its capacity or ability to support people in the future'. According to this definition a reduction of, for instance, the natural resource base can be compatible with safeguarding the population supporting capacity as long as the environmental 'damage' is not irreversible.

Although not explicitly taken into consideration in both the above definitions, applying the carrying capacity concept to humans is also complicated by the fact that per capita natural resource consumption is often extremely variable, both within the same society or among different societies competing for the same natural resources. This makes the carrying capacity a function of distributional elements both nationally and internationally, which adds to the concept's complexity.

Clearly the carrying capacity concept should be defined in a broad and comprehensive manner. Indeed, carrying capacity includes food and energy supplies, ecosystem services (such as provision of fresh water and recycling of nutrients), human capital, people's lifestyles, social institutions, political structures and cultural constraints – all of which interact with one another. Moreover the concept can be specified at various scale levels: at the global level, but also at the national or even subnational level. However, as soon as the global level is left and one turns into the national (or subnational) level, one should realize that the concept for other reasons all of a sudden becomes considerably more complicated. For at a less than global aggregation level one also has to take into account the possibility of international exchange and its impact on the carrying capacity, and thus of economic factors such as international trade and the terms of trade, international financial flows, including investment, and the role of debt burdens.

The technology factor

If the above identity makes one thing clear, however, it is that technology can play a crucial role in counteracting the population pressure on the environment. A classic issue is to what extent technology adapts to population pressure, particularly in developing countries. There is definitely evidence that in the past high population density in developing countries usually has led to a more intense use of farmland, which makes agricultural technology an endogenous rather than an exogenous variable (Boserup). The issue, however, is eventually an empirical one, namely how precisely population growth figures in the various regions relate to technological change and under what circumstances technological improvement does keep pace with the expansion of the population.

The latter is what happened in most of Europe and Asia in pre-modern times. Populations grew slowly. There was plenty of time for technological change. In modern times, the green revolution for a long period enabled food production to keep pace with rapid population growth in most of Asia. However, there are areas and parts of the world where the Boserup thesis did not work and where technology of intensification has not changed in line with population growth. Most of sub-Saharan Africa falls into this category.

A similar incongruence seems to apply to a large extent to soil conservation: conservation technology change does not usually keep pace with population growth, even if land reform is applied. In most places this means loss of forests and woodland. After all, it is cheaper, in the short run, to exploit a resource base beyond its carrying capacity than it is to expand the carrying capacity or to limit population growth. The problem is further aggravated by the fact that growing populations also need extra land for non-agricultural uses, such as housing, roads, factories, offices, playgrounds and parks. Since towns usually grow in agricultural areas rather than deserts or forests, this 'other' land is taken away from farmland. Farmers must either intensify further, or make up their losses by clearing more woodland and forest.

The question therefore, becomes what 'solution' the technology factor could offer, for unlike other species, human beings can expand the carrying capacity of their environment by using technological innovation and trade. However, humans can also diminish the carrying capacity of a region through various forms of environmental mismanagement (e.g. in response to rapid population growth) leading to long-term natural resource degradation.

Technology can increase the carrying capacity of a given region in two ways. First it can allow people to substitute, to some limited extent, a natural resource that is abundant for one that is scarce. Second, technology can increase the efficiency of conversion of natural resources into economic goods, thereby allowing people to 'squeeze' more economic value from a given resource base. So, technological advances can expand the carrying capacity of a region to a considerable extent or, according to some, perhaps infinitely even if the various physical laws e.g. of thermodynamics are respected. (The debate among economists whether or not endogenous technological advance can expand the carrying capacity infinitely is beyond the scope of this study.)

If the analysis applies to the non-global level, the possibility of international exchange can equally expand local carrying capacity by exchanging resources that are locally plentiful for those that are locally scarce. This mechanism only applies under certain conditions, however. The resource that is scarce in one region must be available in surplus elsewhere, and vice versa. There should be no prohibitive obstacles to international trade, e.g. through the cost of transportation or through protectionism, and no internal obstacles in marketing the surpluses abroad in exchange for other values. One should recognize that in practice the above three conditions seldomly are fulfilled for the masses of landless people in the tropical developing regions: the option of international exchange either directly or indirectly has its boundaries.

Population pressure and environmental damage

The present demographic situation and projected trends for the coming decades show that there will be an intensification of the problems associated with rapid population growth such as migration, imbalances in population distribution and rapid urbanization. In general one can argue that countries that are in the process of positive development generally have low population growth rates (2 per cent per year or lower), while countries trapped in the downward spiral of poverty have a population growth of 3 per cent or more. Often, these countries also face serious problems of environmental degradation (Strong, 1991).

Given the 'demographic momentum' built into population growth processes in the tropical countries, and even allowing for expanded family planning programmes, population projections suggest (see also pp.155–156) that in those tropical countries where economies appear likely to remain primarily agrarian, there will surely be progressive pressures on remaining forests, extending for decades into the future (Myers, 1991, p.22).

The adverse environmental impact of population pressure will show itself in various ways, deforestation being one of them. Degradation processes will not always develop smoothly, however. Past events have shown that under certain circumstances and coupled with the burden of past over-exploitation, environmental degradation and natural resource depletion can suddenly and sharply worsen as a result of the demands of growing numbers of people. In this connection ecologists speak of a 'jump effect' of environmental discontinuity. This occurs when ecosystems absorb stress over long periods without much outward sign of damage, but eventually are pushed to the limits of their resilience. They then have reached a disruption level at which the cumulative consequences of stress finally reveal themselves. This is important because it has become increasingly apparent that a population/environment downspiral has been building up for decades in the agricultural sector in the tropical areas.

The resulting non-linear relationships between resource exploitation and population growth seem to relate to various natural resource stocks. The discontinuities can be generally expected to become more frequent, however, firstly due to the demographic momentum and its consequent growing demand on resource stocks (even if fertility rates came down instantaneously everywhere to levels of mere replacement, the world population will keep growing, mainly in the developing regions, for some decades to a level of at least 7 billion), and secondly due to the backlog of concealed costs resulting from past exploitation.

The depletion of fuelwood stocks may serve as an example (derived from Myers, 1992, p.117). Most people in the developing world derive their energy from fuelwood. Already 1.4 billion people obtain most of their fuelwood by cutting it faster than it is being replenished through natural growth, and 130m people cannot meet even minimum needs without over-harvesting stocks. By the year 2000 it is projected that these figures will rise to 2.4 billion and 350m people respectively. Under such circumstances a point is reached where the tree stock quite suddenly starts to decline, albeit gradually at first. Season by season the self-renewing capacity becomes ever-more depleted: while the exploitation load remains the same, the resource dwindles faster and faster. A downward spiral is created and it proceeds to tighten irrespective of the level of exploitation, even if the number of collectors stopped growing (let alone if the number of collectors kept expanding).

Development and the demand for children

The former analysis cannot but lead to the conclusion that population policy is amongst the most structural of measures to combat environmental decline in the tropical regions. Much has been said already about its potential that need not be repeated in this framework; instead we wil try to relate the issue to tropical deforestation as much as possible.

The first aspect that has drawn serious attention in the literature on population relates to the direction of causality between poverty and population growth. Either poverty is one of the main causes of the rapid population growth, or the causality runs the opposite way: poverty is the result of the population explosion. Although various opinions can be found in the literature, the mainstream thinking in the field tends to emphasize the impact of poverty on population increase rather than the opposite.

The analysis on this relationship should ultimately focus at the micro or household level: the question of parental demand for children. One could approach this issue from the cost-benefit perspective (Dasgupta, 1992). The cost side of the ledger in bearing and rearing children is self-evident. Pregnancy involves foregone work-capacity for women, and can involve a considerable addition to her risk of dying. Raising children involves additional costs.

The benefit side is more delicate. One type of motivation stems from a regard for children as children. Not only are children desirable in themselves, they carry on the family line or lineage, and they are the clearest line open to what one may call self-transcendence. Also human beings are probably genetically programmed to want and to value them.

The second kind of motivation stems first of all from the simple fact that in poor countries children are needed on a daily basis for the household. Various numbers of hours a day may be required for obtaining only the bare-essential firewood, dung and fodder. Children can be of great help to relieve the burden of these tasks. In addition, and more importantly, children have an important role to play in the old-age security in an economic environment without any other such provisions. The transfer of resources over a life cycle in poor households in poor countries is usually and in the aggregate from offspring to their parents. This is in sharp contrast to advanced industrial nations, where resources are transferred on average from the adult to the young. Some have concluded on the basis of this difference that whether a society has made the so-called demographic transition, that is the transition from high to low fertility, is among other things related to precisely this direction of the intergenerational flow of resources.

All the above factors relate to the high fertility and low literacy rates in rural areas of most poor countries, and their subsequent impact on environmental degradation.

From a macro point of view one may wonder if a collection of reasoned decisions at the individual level will lead to an outcome which is also optimal at the collective level. The clear linkage between population pressure and environmental decline such as tropical deforestation clearly suggests that the answer on the above question is negative. What explains the dissonance between household and societal levels of decision-making in the field of procreation? From an economic point of view the explanation is related to externalities, i.e. an impact on others, either positive or negative, that cannot be charged upon the person who ultimately is responsible for this impact.

Two sources of externalities present themselves in relation to this aspect of the population issue. The first has to do with the fact that although increased population size implies greater or even unacceptable environmental pressure (defined in its broadest sense), one would not expect households, acting on their own, to take account of the fact that the children they will have will affect others by adding to this pressure. The second source of externalities is related to the fact that procreation is not only a private matter, but a social activity as well. In many societies there are norms encouraging high fertility rates which no household desires unilaterally to break. Such norms often survive even when their purpose or economic rationale has disappeared. As a result human behaviour that is perfectly rational from the social point of view of an individual household, can at the same time pose a major problem at the community level.

Population size and population growth

Demographic analysis based on both time-series and cross-section data suggests a clear inverse but declining relationship between fertility (and mortality rates) and national income per head (for some recent empirical evidence, see e.g. Dasgupta, 1992). The question that has come up recently is to what extent fertility rates in the developing world can be linked with environmental decline. With respect to this relationship, the causality could again run either way. On the one hand one finds the 'classical' pattern of causality: high fertility rates are to a large extent responsible for high rates of population increase, and thus for the consequent environmental degradation.

On the other hand, among other things, for poor rural households environmental degradation could also be a cause of increased family size. Each household needs many hands, and it can be that the overall usefulness of each additional hand increases with declining resource availability. If the latter hypothesis (not yet systematically tested) is correct one notices another adverse vicious circle (Dasgupta, 1992, p.96): a high rate of fertility and population growth further damages the environmental resource base, which in turn in a wide range of circumstances provides further incentives for larger families, which in turn further damages the resource base and so on.

In view of the many uncertainties with respect to the validity of the final hypothesis, most studies focus on the opposite relationship: how does population growth relate to the environment? This relation is expressed in the initial identity at the beginning of this section. But, if one focuses on the population element it is clear that in establishing this relationship, a distinction should be made between the impact of the absolute population size of countries or regions within countries on their natural resource base on the one hand, and that of the rate of population growth on the other hand. Both factors can in their own way affect the environment adversely. The distinction is especially relevant because it helps us to understand why some countries and regions, even those with relatively low population densities, are yet confronted with severe environmental dislocations. What seems to matter is that population size and population density of countries are not per se the causes of problems such as natural resource degradation or even hunger. Rather, these problems arise when the population expands too fast in relation to the productivity of the resource base upon which it survives. Parts of sub–Saharan Africa provide perhaps the clearest examples of how much of a seemingly 'underpopulated' area may, in fact, be too crowded, but also that the effects of rapid population increase are always mediated by such factors as land-tenure arrangements, natural resources availability, and technology. Clearly, a rational manipulation of the latter variables via public policy and planning can under those circumstances be used to achieve more efficient land-use responses, to successfully accommodate higher populations.

Population policies

Total fertility rate is the number of live births a woman would expect to give were she to live through her childbearing years and to bear children at each age in accordance with the prevailing age-specific fertility rates. The measure pertains to the number of live births, not pregnancies.

All societies practice some form of birth control: fertility is below the maximum possible in all societies. However, concern about rapid population growth in the 1950s and 1960s led to the implementation of family-planning programmes, often with donor assistance, in several developing countries, and the number of such programmes has expanded rapidly throughout the 1970s and the 1980s. These programmes vary widely, from relatively low-key operations that attempt to provide access to modern family planning methods for a wide spectrum of couples, to more active interventions through subsidies, incentives, and in some cases even social arm-twisting.

Meanwhile fertility has fallen sharply in some parts of the world (Hill, 1992), particularly in East Asia, where it has reached levels consistent with long-term stabilization of population numbers. Decline has been slowest in sub–Saharan Africa, where until the last five years or so there was no evidence of any sustained decline in fertility. However, it is now clear that fertility is also falling in some parts of Eastern and Southern Africa, though it remains high in many parts of the region.

Various factors explain why fertility rates not only are systematically higher in poorer countries, but also can be reduced less easily.

1) There is a strong inverse relationship between the level of education, particularly for women, and fertility rates. This has long been empirically documented.

   The depressor effect of female education on fertility operates both through increased use of family planning and through improved health and hygiene practices, lowering infant mortality. It is difficult to imagine further significant decreases in worldwide fertility rates in the absence of progress in increasing overall access in education, and furthermore, in narrowing the male-female literacy gap . . . . Not only is basic education important, but access to specialized information, namely on contraception, is critical. And hundreds of thousands of poor people suffer today from lack of access to such information.

   (Ferguson-Bisson, 1992, p.93)

   The relationship between the mothers' education and their fertility rates is further illustrated by the data in Table 6.2.

2) Related to the first argument, the impact of family-planning programmes appears to be greatest in countries with higher levels of development, and thus weaker demand for large families (Hill, 1992). Family-planning programmes have clearly contributed to the worldwide fertility decline, largely on the supply side through satisfying unmet need for family planning, but also to some extent by influencing demand for children. However, in traditional societies with high demand for large families, the potential for quick results from family-planning programmes appears rather limited.

   The latter has recently been confirmed empirically by Bongaarts, Mauldin, and Phillips, 1990. Based on a cross-national regression analysis of fertility levels on intervention programme variables, they concluded that, although programme effort per se was not

Table 6.2 Education and fertility rates – demographic characteristics

	Fertility rate % 1985–90	No. of children[1] by mother's years of education		Contraceptive use %[2]	Unmet need est. %[3]	
		none	>6		high	low
Latin America						
Bolivia	6.1	na	na	26	24.2	6.8
Brazil	3.5	na	na	66	na	na
Colombia	3.6	7.0	2.6	65	na	na
Ecuador	4.7	7.8	2.7	44.2	25.8	12.9
Guyana	2.8	6.6	4.8	31.4	28.5	23.3
Peru	4.5	7.3	3.3	45.8	41.1	13.9
Surinam	3.0	na	na	na	na	na
Venezuela	3.8	7.0	2.6	49.3	21.5	10.4
Africa						
Angola	6.4	na	na	na	na	na
Benin	7.0	7.4	4.3	9.2	na	na
Cameroon	5.8	6.4	5.2	2.4	1.1	1.0
Centr. Afr. Rep.	5.9	na	na	na	na	na
Congo	6.0	na	na	na	na	na
Côte d'Ivoire	7.4	7.4	5.8	2.9	3.0	1.8
Eq. Guinea	5.7	na	na	na	na	na
Gabon	5.0	na	na	na	na	na
Ghana	6.4	6.8	5.5	9.5	7.5	5.9
Guinea	6.2	na	na	na	na	na
Guinea–Buissau	5.4	na	na	na	na	na
Kenya	8.1	8.3	7.3	17	9.9	na
Liberia	6.5	na	na	6.5	na	na
Madagascar	6.6	na	na	na	na	na
Nigeria	7.0	6.6	4.2	4.8	na	na
Senegal	6.4	7.3	4.5	11.7	na	na
Sierra Leone	6.5	na	na	na	na	na
Togo	6.1	na	na	na	na	na
Uganda	6.9	na	na	na	na	na
Tanzania	7.1	na	na	na	na	na
Zaire	6.1	na	na	na	na	na
Asia						
Indonesia	3.3	na	na	48	15.3	10
Malaysia	3.5	5.3	3.2	51.4	22.6	14.7
Philippines	4.3	5.4	3.8	45	29	11.1
Thailand	2.6	na	na	66	17.3	11.1
Kampuchea	4.7	na	na	na	na	na
Lao Dem. Rep.	5.7	na	na	na	na	na
Vietnam	4.1	na	na	20	na	na
Myanmar	4.0	na	na	na	na	na
Fiji	3.2	na	na	41	na	na
Papua New Guinea	5.7	na	na	na	na	na
Solomon Isl.	na	na	na	na	na	na

Sources: WRI (1990), Tables 16.2 and 16.5, and Boulier (1985), Table 1.

Notes: 1) For women who have completed less than one year of school and those with seven or more years of education (1985 figure). In general, the latter marry nearly four years later, have approximately 25 per cent higher contraceptive use, and breastfeed their babies for a shorter period. Education is also associated with lower maternal and child mortality. For this reason education

may not always have a net negative effect on numbers of children who are
born or who survive.
2) Contraceptive prevalence is the percentage of couples using contraception
when the woman is of childbearing age.
3) The high estimate of unmet need is defined as the percentage of women
wanting no more children, who are not pregnant, infertile, or using
contraceptive methods. The high estimate assumes that lactating women and
users of inefficient methods have the same unmet need as women wanting
no more children who are exposed to the risk of pregnancy and are not using
any method of contraception. The low estimate assumes that, among women
who want no more children, those who are breastfeeding for less than one
year or using inefficient methods would neither demand nor benefit from
modern contraceptive methods were they more readily available. Figures
relate to the period 1975–1981.

found to be significantly related to TFR, the interaction between
programme effort and the development level was. They found that
in a country with a high level of development, an increase in
programme effort from 'very weak' to 'strong' would be associated
with a reduction of 2.3 births per woman, whereas in a country
with a low level of development, the same increase in programme
effort would be associated with a decline of only 1.1 birth per
woman.
3) The use of contraceptives has for one reason or the other been
uneven within poor countries. The large variations that occur
across regions not only reflect a divergence in the public provision
of family planning and health-care services, they reflect variations
in demand as well. Surveys indicate that woman themselves
perceive an unmet need for access to methods to reduce their
fertility.

THE RURAL VERSUS THE URBAN POPULATION

After having discussed the relationship between population growth
and environmental decline in general, we will now turn to the issue of
to what extent one should distinguish various groups within the total
population whose impact on deforestation may differ significantly. A
first distinction, corresponding with a major part of the literature on
the issue, is between the rural and the urban part of the population in
the tropical countries (some primary figures have already been
presented at the beginning of this chapter). For only part of the rural
population can be held directly responsible for deforestation; the urban
population can only indirectly contribute to it. Consequently any
developments determining the (relative) size of the rural population,
such as rural-urban and urban-rural migration, can eventually have a
decisive impact on deforestation.

In nearly all developing countries in the world a strong tendency
towards urbanization has come to the fore during the last decades; both
in absolute and in relative terms more and more people are living in
urban agglomerations. In 1950 the share of the world's population

living in urban areas was only 29 per cent; in 1990 it was already 50 per cent (World Bank, 1992, Table 31). Projections made by the United Nations indicate that more than 60 per cent of the world population will live in urban areas around the year 2025.

Rate of urbanization: parameter to measure the intensity of the redistribution of population from rural to urban areas; the average annual rate of increase in the per cent urban. A positive rate of urbanization indicates that the urban population is increasing at a faster pace than the total population. In the less developed regions, the rate of urbanization was 1.4 per cent during the 1970s and 1.5 per cent during the 1980s. According to the present projections, the rate of urbanization in the less developed regions will peak at 1.7 per cent for the period 2000–2005, before declining steadily to 1.2 per cent for the period 2020–2025. These figures indicate that the rate of population redistribution from rural to urban areas in the less developed regions will remain strong for the rest of this century.

(UN, 1989)

The above-mentioned figures do not reveal the large differences in urbanization trends between the major regions in the world. Some data may serve to illustrate their wide variety. In 1990, 26 per cent of the South Asians lived in urban areas, 29 per cent of the sub–Saharan Africans, in East Asia and the Pacific 50 per cent, in Europe 60 per cent, in Latin America and the Caribbean area 71 per cent, and in the United States 75 per cent (World Bank, 1992, Table 31). Past trends of urbanization figures in the tropical regions (see Table 6.3) show, however, that Asia and Africa, presently at the low end of the range, are in a process of 'catching up' with the more developed regions where urbanization rates commonly fluctuate around 70 per cent. Moreover, most projections (see also Table 6.3a on p.154 and Table 7.4 on p.202) point out that in the tropical regions these processes will keep going for a considerable period of time to come. It therefore seems fair to presume that urbanization will increasingly become a key problem in the tropical developing world with as yet still unforeseen environmental consequences.

Urbanization rates of countries that still have large areas of tropical rainforest are presented in Table 6.3a. Table 6.3a clearly illustrates the dramatic increase in the urbanization rates in the various tropical countries during the past four decades. In addition, just as expressed in Table 6.1, the large differences in urbanization rates between the three main tropical regions comes to the fore. Urbanization rates in the Latin American tropical countries exceed by far those of the other two tropical regions, Asia and Africa. Even Bolivia, in Latin America among the countries with the lowest rate of urbanization, has a larger urbanization rate than any other country on the list in tropical Africa or in tropical Asia.

The urbanization patterns gradually have somewhat shifted during the past in a different way among the main tropical regions. In 1950, most of the African countries for instance had rather low rates of urbanization, compared to those of the Asian countries. In 1990 the

Table 6.3 Urbanization rates in the tropical regions

a) *Urbanization rate (%), 1950–1985*

	1950	1955	1960	1965	1970	1975	1980	1985
Tropical Asia								
Myanmar	16.1	17.6	19.3	21.0	22.8	23.9	23.9	23.9
Indonesia	12.4	13.5	14.6	15.8	17.1	19.4	22.2	25.3
Lao Dem. Rep.	7.2	7.6	8.0	8.3	9.6	11.4	13.4	15.9
Malaysia	20.4	23.4	25.2	26.1	27.0	30.5	34.3	38.2
Papua New Guinea	0.7	1.4	2.7	5.2	9.8	11.9	13.1	14.3
Philippines	27.1	28.7	30.3	31.6	33.0	35.6	37.4	39.6
Thailand	10.5	11.5	12.5	12.9	13.3	15.2	17.3	19.8
Vietnam	11.7	13.1	14.7	16.4	18.3	18.8	19.3	20.3
Tropical Africa								
Angola	7.6	8.9	10.4	12.5	15.0	17.8	21.0	24.5
Benin	6.6	8.0	9.5	11.4	16.0	21.5	28.2	35.2
Cameroon	9.8	11.7	13.9	16.4	20.3	26.9	34.7	42.4
Centr. African Rep.	16.0	19.1	22.7	26.7	30.4	34.2	38.2	42.4
Côte d'Ivoire	13.2	16.0	19.3	23.1	27.4	32.2	37.1	42.0
Ghana	14.5	18.5	23.3	26.1	29.0	29.8	30.7	31.5
Guinea	5.5	8.4	9.9	11.7	13.9	16.3	19.1	22.2
Kenya	5.6	6.4	7.4	8.6	10.3	12.9	16.1	19.7
Liberia	13.0	15.6	18.6	22.1	26.0	30.4	34.9	39.5
Madagascar	7.8	9.1	10.6	12.3	14.1	16.3	18.9	21.8
Nigeria	10.1	12.1	14.4	17.0	20.0	23.4	27.1	31.1
Senegal	30.5	31.2	31.9	32.7	33.4	34.2	35.0	36.4
Sierra Leone	9.2	11.0	13.0	15.4	18.1	21.1	24.6	28.3
Togo	7.2	8.4	9.8	11.3	13.1	15.8	18.8	22.1
Uganda	3.1	4.0	5.1	6.5	8.0	8.3	8.7	9.4
Zaire	19.1	20.7	22.3	26.1	30.3	32.2	34.2	36.6
Tropical Latin America								
Bolivia	37.8	38.5	39.3	40.0	40.8	41.5	44.3	47.8
Brazil	36.0	40.4	44.9	50.4	55.8	61.8	67.5	72.7
Colombia	37.1	42.6	48.2	53.5	57.2	60.8	64.2	67.4
Ecuador	28.3	31.3	34.4	37.2	39.5	42.4	47.3	52.3
Peru	35.5	40.8	46.3	51.9	57.4	61.4	64.5	67.4
Venezuela	53.2	60.1	66.6	69.8	72.4	77.8	83.3	87.6

Source: United Nations, Prospects of World Urbanization 1988, New York, 1989.

picture has reversed, because between 1950 and 1990 the rates of urbanization in the African tropical countries generally more than tripled, whereas those of the Asian tropical countries generally 'only' doubled.

In the Latin American tropical countries during the same period the share of people living in urban areas generally increased some 30 percentage points (except for Bolivia with an increase of 10 percentage points only). This generally contributed to the already very high urbanization rates. In the 1980s, Venezuela reached an urbanization rate of over 80 per cent! In Latin America, the countries having the largest increase in urbanization were Mexico, Colombia and Brazil. These were also the countries with the strongest growth in their economies (WRI, 1990, p.66).

Although the African and Asian countries are in a process of rapid

b) Urbanization (%), projections

	2000	*2010*	*2020*	*2025*
Tropical Asia				
Myanmar	28	35	43	47
Indonesia	36	44	52	56
Lao Dem. Rep.	25	33	41	45
Malaysia	50	58	64	67
Papua New Guinea	20	27	34	38
Philippines	49	56	63	66
Thailand	29	37	45	49
Vietnam	27	35	43	47
Tropical Africa				
Angola	36	44	52	56
Benin	53	60	66	69
Cameroon	60	66	72	74
Centr. African Rep.	55	61	67	70
Côte d'Ivoire	55	61	67	70
Ghana	38	45	53	57
Guinea	33	41	49	53
Kenya	32	40	48	51
Liberia	52	59	66	68
Madagascar	32	40	48	52
Nigeria	43	51	58	62
Senegal	44	52	59	63
Sierra Leone	40	48	56	59
Togo	33	41	49	53
Uganda	14	19	26	30
Zaire	46	54	61	64
Tropical Latin America				
Bolivia	58	65	70	73
Brazil	83	86	88	89
Colombia	75	79	82	84
Ecuador	65	71	75	77
Peru	75	79	83	84
Venezuela	94	95	96	96

Source: United Nations, Prospects of World Urbanization 1988, New York, 1989.

urbanization, their urbanization rates are still well below that of the more developed regions. However, it is very likely that this gap will decrease substantially in the next few decades.

UNDP (1990, pp.85–95) has pointed out that the rapidly increasing concentration of people in cities is by now almost exclusively a developing country phenomenon. The overall urban population of the developing countries, now 1.3 billion, is expected to grow by nearly another billion in the next 15 years. This is reconfirmed by the rather authoritative United Nations projections of urbanization rates. These projections are based on the so-called 'United Nations method' (UN, 1980), basically involving an extrapolation of the most recently observed pattern of urban/rural growth difference. This method results in a logistic time path of the proportion urban which has a peak velocity (annual absolute gain in proportion urban) at a proportion of 0.5 and has a maximum urban proportion, eventually reached by all countries, of 1.0 (for more details on the UN methodology and our criticisms, refer to pp.195-199).

According to the UN projections, the total developing world's rural population will reach an upper limit by the year 2015; beyond this point all future population growth will be concentrated in urban areas. Consequently by the year 2015 half the developing world's people are expected to live in urban areas. It should be mentioned, however, that the UN method does not explicitly introduce the element of the absorption capacity of cities. If it turns out in the future that there are certain limitations to the degree of congestion of major cities, one could imagine that population pressure gradually decreases as people move back to the traditional rural areas. This could then have a rather decisive impact on deforestation processes.

As was mentioned earlier, most projections indicate that the trend towards further urbanization in the tropical regions will also continue during the next decades. The UN projections with respect to urbanization for the period up till 2025 for the tropical countries in particular have been represented in Table 6.3b (here the projections are based on 1985 as the final year with real urbanization data available).

Table 6.3b makes clear that the considerable shift towards urbanization in the tropical areas during the past decades will continue for the next few decades to come with a comparable speed. According to the projections by 2025 the urbanization rate in virtually all tropical countries will surpass 50 per cent, even in those countries that can now (and even then) be considered sparsely populated. In most countries the rate varies by then between 50 and 70 per cent, except for tropical Latin America where rates almost always will be higher than 75 per cent.

Most of this growth – two-thirds of it in many Asian and Latin American cities – will be the natural increase of populations already in the cities. The remainder will come from rural-urban migration, the incorporation of villages into expanding urban municipalities and the changing definitions of settlements from rural to urban as they reach a given size (UNDP, 1990, pp.85-86).

It speaks for itself that the contribution of rural-urban migration to the explanation of the rapidly increasing urbanization rates in the tropical areas differs across the various countries. In Indonesia, to give just an example, intra-provincial migration used to be relatively rare (World Bank, 1986). Less than 7 per cent of the population was living outside their province of birth, of which half had migrated to urban areas. In the last decade intra-provincial migration seems to be accelerating, however, rural-urban migration being its most important source. Consequently about half of Indonesia's increase in urban population during the last decade (total 9.6 m people) was due to migration: during the same period 4.6 m people migrated from rural to urban areas; the natural increase of the urban population provided the other half (World Bank, 1986, Table 4).

THE RURAL–URBAN MIGRATION ISSUE

Various explanatory factors have been put forward in the literature relating to the substantial net rural-urban migration in the developing countries. The UN, in one of its past studies (1980, pp.30–31) mentioned the following basically economic determinants:

1) **Unbalanced technological change.** Given a certain rate of technological and productivity growth in non-agricultural activities, the faster the productivity growth in agricultural activities, the faster in general will be movement out of the agricultural sector.

2) **International economic relations.** Concentration of exports in one sector and of imports in another can attenuate or promote the internal structural transformations that typically produce urbanization.

3) **Population growth rates.** High rates of natural increase in rural and urban areas can change relationships among factors of production. It appears to be commonly assumed that diminishing returns to labour are more prominent in rural than in urban areas because of limited supplies of land. In this case, more rapid rates of natural increase, even though equal in rural and urban areas, would depress rural incomes more than urban and lead to accelerated urbanization.

In addition the major impact of the institutional setting and government interference on rural-urban migration was emphasized:

4) **Institutional arrangements governing relations among factors of production.** A land tenure system that reduces the absorptive capacity of rural areas may accentuate the response to population growth. So may financial systems that make capital formation easier in urban than in rural areas. The exclusionary land tenure system in much of Latin America is often cited as an important factor in its rapid rural-urban migration. In a more general sense, price and tax distortions that discriminate against rural areas are pervasive and probably foster migration towards urban areas.

5) **Biases in government services.** It is clear that health and educational expenditures in the less developed countries are directed towards urban areas in disproportionate amounts to urban population size. It is not always so clear that this pattern should be termed a bias, since it may reflect the greater cost effectiveness of urban expenditures in these as well as in other services. That is, the tendency may reflect agglomerative economies. To the extent that true bias creeps in, however, urbanization is independently accelerated.

6) **Government policies on migration.** To the extent that these policies are successful in deflecting the course of events, rural-urban migration will be slowed.

The importance of economic motives in explaining migration patterns in developing countries is also reflected by levels of urbanization by age and sex. In the less developed regions urbanization levels generally begin to rise beginning with the age group of 10-14 years. In developing countries people start to migrate at a younger age than in the developed countries due to the younger accession to the labour force. Another motive for this relatively young age at which people start to migrate is that the educational opportunities are generally better in urban areas. In developing countries, the proportion living in urban areas reaches its peak in the age group 20-29.

However, often people start to return to rural areas after having succeeded in their original motive for migrating to the urban area. The attachments to place of birth or current residence are no doubt powerful factors. Family building and retirement also play an important role. These factors definitely affect the nature of rural-urban migration (short-term, family in stages, high turnover), but probably far less its long-term volume.

If one searches for the deeper policy forces that have contributed to the historical process of rural-urban migration in the developing countries, one encounters the fact that from as early as the 1930s in many developing countries there was a shift of emphasis from export crop production to technologically-dependent industrial production, as well as from local staples to export crops and livestock. Economic surplus from primary production was channelled into technologically-advanced production in a few urban centres, accentuating a form of economic 'dualism'.

A contributing factor and an immediate consequence was economic stagnation among small rural producers, putting them under strong pressure to migrate into the cities. This was reflected in growing landlessness, as tenants smallholders were dispossessed of land to make way for large-scale commercial farming, and as land fragmentation proceeded with population growth and curtailed access to cultivable land, and as agricultural involution made family smallholdings insufficiently large to permit even a bare survival subsistence from farming.

Changes in the terms of trade between staple food crops and export crops, between agricultural and urban-industrial goods, and between local and imported goods, also worsened and added pressure on the rural poor to seek alternative livelihoods. In more and more cases, the delicate process by which the peasantry had for generations reproduced themselves and their source of livelihood came under such strain that long-term emigration by youthful members of rural households increasingly replaced the circulation that had more traditionally been a means of temporary adjustment to life-cycle needs and periodic stress. Circulation, notably seasonal movements, gave way in numerous places to long-term structural migration, much of it to urban areas . . ..

(Standing, 1984)

The migration patterns differed among the major regions, however.

- **In Latin America**, the rural-urban migration during the last decades was strongly related to the rapid commercialization of agriculture and to the actual patterns of land reform. Despite the potential of land reform to arrest rural out-migration, in reality most Latin American redistribution schemes' main impact was through accelerating the transition to commercial farming by landlords. This often caused tenants to be pushed off the land as landowners converted to agro-export crops or expanded livestock production. In addition attached labourers have been widely replaced by use of seasonal and casual labour. In effect, according to Standing (1984), land reform as practised in countries such as Bolivia, Chile and Peru, helped increase rural out-migration.

The commercialization of agriculture in Latin America and the transition of social relations of production that accompanied land reform increased the seasonality of employment and led to a growth of labour circulation, often from the peasant-occupied highland areas to valley estates. The quantitative impact of such labour circulation on deforestation is as yet unknown.

- **In Africa**, originally urbanization was checked not only by the temporary nature of much of the movement but also by the fact that mortality rates in urban areas were extremely high long into the twentieth century. More recently, however, urbanization rates are increasing rapidly. An increasingly important aspect of African rural-urban migration has become the altered pattern of labour circulation. Indeed, men have long constituted the great majority of seasonal and other circulants to the cities, resulting in separation of husbands and wives for large periods of time in which childbearing is at its height. In recent years, however, growing proportions of adult male migrants have become more rooted to urban areas, staying there continuously for years rather than months at a time. To compensate, there has been a growth in female circulation, as wives have gone to the towns to stay with their husbands or to find temporary work for themselves. So, increased duration of male out-migration has increased female mobility, and probably raised fertility levels (Standing, 1984).

At the same time population growth in settled rural areas in Africa has led to various 'spontaneous resettlements' by agricultural groups, taking over land periodically used by nomads, or even by shifting cultivators, thus disrupting the subsistence strategy of such migratory populations.

- **In Asia**, particularly in rural areas in South and Southeast Asia, the pace of agrarian change has not been rapid, and in such places the proportions leaving the villages permanently have not been large. Limitations on the absorption capacity of great cities have also checked rural-urban migration, while some efforts in the region have been devoted to urban decentralisation and rural resettlement programmes. Given the nature of industrialization processes in many of the so-called Newly Industrializing Countries (NICs), the cities have attracted at least as many young women as men migrants. Another feature of population mobility in Asian countries is labour circulation, a reflection of land fragmentation and the agrarian structure of production. But even so, just as in the other tropical regions the duration of circulation has grown in recent years.

Migration policies

One could argue first that developing countries' governments during the past decades implicitly favoured urban development through their preferential treatment of cities for industrial development, for infrastructure investments, for social services and for food and other subsidies. Secondly, government policies to reduce or even revert rural-urban migration generally were not very successful. In fact integrated rural development schemes to raise agricultural productivity and persuade people to remain on the farm were often overly complex and lacked the necessary manpower for effective implementation, especially in Africa.

Colonization projects opening new lands for settlement, besides being rather costly, have in practice usually only benefited a small proportion of the rural poor, and therefore could not seriously restrain rural-urban migration. Such projects have tended to be implemented where social tensions have risen because peasants have been under economic pressure or because landlords have been unable to secure adequate surplus, but above all as an alternative to land reform. In fact after the introduction of colonization schemes in many cases large proportions of the settlers have left.

Another rather ineffective measure to restrict urban growth has been the setting of minimum rural wages to reduce rural-urban wage differences, a major incentive to migration. But where this measure has been implemented, if not circumvented it supplanted permanent labourers with seasonal workers and accelerated farm mechanization. The minimum rural wage thus led to rural unemployment and declining income, probably stimulating rural-urban migration.

The demolition of squatter settlements, undertaken for many years in large parts of Latin America, Asia and Africa, was also not really successful. These highly unpopular policies usually focused on the destruction of new squatter settlements, while tolerating older squatter settlements, and have had little permanent effect. Indonesian officials tried in the early 1970s to regulate migration to Jakarta by issuing temporary permits requiring new migrants to find housing and employment in six months or face deportation. This resulted in an enormously profitable pattern of small-scale corruption, as migrants bought time in the city, and they led to a rapid growth in commuting and circulation between surrounding villages and the city. Largely ineffective, the controls were soon discontinued.

7

MODEL INTEGRATION

INTRODUCTION

In the previous chapters various socio-economic factors underlying deforestation in tropical regions have been discussed. In order to model these variables in relation to deforestation in tropical regions a global simulation model, IDIOM, has been developed. IDIOM can be used for scenario analyses such as of the impact of policy options on the tropical forest area. The (modular) structure of IDIOM, and the theoretical considerations underlying the model will be discussed in the following sections.

In Chapter 4 TROPFORM, the model which served as the starting point in the process towards modelling deforestation on a global scale, was presented. One of the shortcomings of TROPFORM is the rather crude mechanism of economic interaction between regions in the world. In TROPFORM, only tropical timber trade and logging activities between regions have been modelled, and it incorporates the need for agricultural land rather crudely. In IDIOM a more extensive interaction between regions has been developed. Since one of the main causes for deforestation is clearance of forest for agricultural purposes, it is necessary to model, for example, trade in agricultural products between regions.

As has already been mentioned in Chapter 1, the main cause of deforestation is the conversion of forest into agricultural land. To build the need for agricultural land into the model structure, and to establish the necessary linkages between the various model components, a land use module was designed. The land use module, and theoretical considerations underlying the modelling of the subsistence and the commercial agricultural sector, will be discussed on pp.162-178.

The main challenge of this project was to build an integrated model in which TROPFORM and the land use model are combined with a module containing variables which depict the more general socio-economic processes underlying deforestation (see Chapter 6). SARUM (Systems Analysis Research Unit, 1978) served this purpose, being a global simulation model which has been used and adapted for several research projects. It deals extensively with economic interactions between regions in the world. SARUM has therefore been integrated with TROPFORM in order to be able to model interactions between different regions more extensively. A description of SARUM is given (on pp.178-183), followed by a separate section on modelling migration.

The final integration of the three IDIOM modules and the database which has been used for the simulation of the scenarios of Part III, are discussed on pp.183–204, and illustrated in Figure 7.11.

LAND USE MODELLING

For our research purpose TROPFORM has been updated, programmed in a user-friendly way, slightly amended, and tested. The model was subsequently applied to the Indonesian case (see also Appendix 2). This model constitutes the first element of IDIOM, and focuses on supply, demand and international trade of tropical timber; its deforestation module, relating deforestation to land use, is however, rather crude.

SARUM (see pp.206–208) was also updated in 1989; its database has been reorganized so as to reflect the regional composition which best satisfies the problem, i.e. to analyse global tropical forest degradation and extinction. The model provides a broad multiregional economic framework which sets the stage for a coherent analysis and allows for an evaluation of the developmental aspects of the issue. This model constitutes a second element of IDIOM.

The third element which is meant to enable a useful linkage between TROPFORM and SARUM is a land use module. The basic philosophy in the timber producing countries is that land use is mainly due to agricultural activities. Given the available land area and given the several agricultural activities carried out on the land and their productivity, a situation may arise wherein additional agricultural activity may necessitate deforestation. To assess the precise mechanisms which may cause deforestation, a complete and thorough description of economic development processes is pertinent.

Indeed, if one thinks of the several economic activities involving direct land use, a distinction should be made on the basis of land productivity. For if the productivity of all land in a region were the same, one could easily establish a relationship between land use and agricultural production, at least in the case of a closed economy. Given the required amount of food per capita and agricultural productivity one can determine how much land is required to feed the population. If the actual development is such that more land needs to be exploited, new land will be developed. Land development costs will subsequently determine whether or not the new land is acquired through deforestation (the speed of exploitation of the forest surpasses its regeneration capacity). In fact, the original TROPFORM model incorporated just such a crude land use mechanism.

In reality, however, land is used by different categories of end-users and for various purposes. Consequently, agricultural productivity varies widely across the users in one and the same region. Moreover, the size of the several agricultural activities is determined by different underlying causal factors. To give an example, the size of commercial cropping will be more sensitive to the pattern of export prices for the cash crops, than will the subsistence agriculture activities. By contrast, the size of subsistence activities will be highly sensitive to the size of the subsistence population, whereas commercial cropping activities are not.

In this section attention will be paid to the main theoretical framework underlying our modelling of tropical land use and deforestation. The main hypothesis is that every developing region needs to produce a minimal amount of food, required for domestic use. Obviously, this amount needs to expand in proportion with the size of the population of the region. One could make the assumption that regions could exchange surpluses and shortages of food among themselves. In the following, however, where we focus on tropical developing regions in particular, the above assumption on the minimal amount of food produced in every region seems justified in view of the large subsistence sector in these regions.

Given the amount of food which needs to be produced in the various tropical regions, the issue becomes how and under what conditions agricultural production takes place. Here the assumption will be made that the increase of agricultural production which is required due to population pressure, can essentially be achieved through two main routes. The emphasis in the first route is on the expansion of the area employed for agricultural purposes. In the tropical areas this often means deforestation. The other route, instead, focuses on the increase of land productivity in the tropics through investments geared towards higher productivity in agriculture.

At the micro level the choice as to whether or not additional land will be 'developed' for agriculture, or if investment is focused on increasing land productivity, is in our model based on economic judgement, given the various legal, political and social constraints. The experience has shown that at the fringe of the tropical forest often deforestation turned out to be the result of this decision-making process. The question is how the economic behaviour in the agricultural activities at the forest fringes can be modelled, what empirical information is available, and eventually how and to what extent the alteration of parameter values can have an impact on human behaviour. Obviously the final goal is to evaluate whether policy variables can be designed in such a way that the farmers will alter their behaviour and deforestation will be halted.

The main trade-off one faces in discussing the deforestation land use nexus is between increasing land productivity and increasing the land area under exploitation. Both options involve private and social costs and benefits. Increasing land productivity involves private costs in terms of agricultural technology inputs, increased fertilizer use, the introduction of agroforestry systems and so on. In addition, the risk of long-term environmental costs due to land overuse should be mentioned; they constitute the potential social costs of this option. The private benefits of increasing agricultural productivity is the subsequent increased production; the main social benefit, less deforestation.

The private costs involved with the other option, clearing the forest for agricultural purposes, are first of all related to the clearance of the land and the need of the farmers to move further from their original sites. The welfare costs are the main subject of the study, deforestation with all its adverse implications both environmentally, economically and socially. It should be clear beforehand that a precise dividing line between welfare and private costs cannot always be drawn easily. Various sources point for instance at the enormous opportunities to exploit the forest sustainably, e.g. by focusing on non-timber tropical products, and at the same time at a much larger economic return than if the same area were cleared and used for agriculture. In such cases deforestation should not only be seen as a welfare cost but also as a private cost. The private benefits of deforestation are that they can provide the farmers with a living; the social benefits could be that the population pressure on the overpopulated cities is somewhat relieved.

In the following model, the essential choice as to whether the required incremental food production will be based on higher land productivity or extended land use, will be analysed as an economic problem, whereby individuals base their decisions on economic incentives.

With respect to increased land productivity, the assumption was made that the law of diminishing marginal returns applies. It says that the physical yield per hectare, for instance expressed in cereal equivalents, increases with the amount of agricultural technology inputs (in the broad sense), but the less so with an increasing amount of technology inputs invested per hectare. (The resulting yield curve is defined at a given level of knowledge about agricultural technology; a rise in the level of knowledge causes the yield curve to shift upward.) Beyond a certain point additional technology investments per hectare could even reduce the yield per hectare. Consequently, the shape of the relationship between both variables could be as presented in Figure 7.1.

The theoretical underpinning of the law of decreasing returns is that the above relationship represents a portion of a broader production

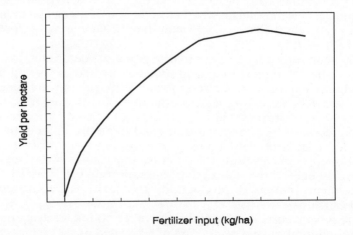

Fertilizer input (kg/ha)

Figure 7.1 Yield function of land

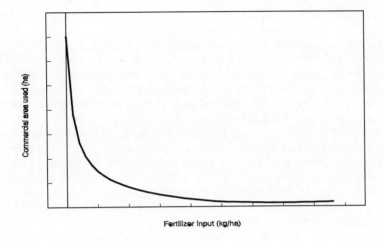

Figure 7.2 Isoproduct curve (output = 1000 kg)

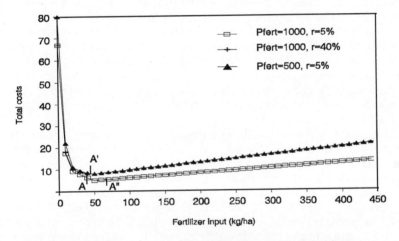

Figure 7.3 Total cost of agricultural technology and land development

function. It is therefore partial and has to rely upon the *ceteris paribus* assumption with respect to the other determinants of the yield per hectare. These could be other economic factors, but also social or political constraints. By assuming these other determinants are constant, they will increasingly act as constraints as the agricultural technology investments per hectare are further increased. This causes the marginal returns to decline and eventually also the absolute returns.

Based on the above yield function, the choice the farmers are facing between increased land productivity and increased land use can be clarified. In Figure 7.2 the yield curve of Figure 7.1 was used to develop an isoproduct curve, representing all combinations of agricultural technology investments per hectare and the size of the area used for agricultural purposes that together can produce a pre-determined agricultural output. If the assumption is made that technology is non-discrete, an iso-product curve could have the following form: a negative slope characterized by a decreasing marginal rate of substitution between the area used and technology input up to a certain point; thereafter a positive slope due to the decreasing absolute returns in the production function. Obviously many iso-product curves could be drawn: the higher the amount of output they represent, the more the curve shifts away from the origin.

The crucial issue now has become on what point of the iso-product curve the individual farmer reaches his optimum: this determines the size of the deforestation. Clearly the marginal costs involved with introducing the agricultural technology will be weighted against those of expanding the area. Given the amount of food to be produced, the farmer will try to find out which combination of both options will lead to the minimal amount of total costs.

The private costs involved with realizing increased agricultural productivity can be determined on the basis of the aforementioned production function (in theory an excessive technology investment per hectare can lead to adverse environmental effects, for instance toxification and salinization through fertilizer overuse; we assume that these social costs generally are non-relevant in the tropical areas under consideration). If more technology is invested per hectare, less hectares will be needed for achieving the required production level. Due to the decreasing marginal returns, the reduced size of the area cannot compensate for the increased technology intensity, however. The total costs involved with agricultural technology investments will therefore rise with this technology intensity.

The costs involved with expanding the land area are, instead, negatively related to the technology intensity. This is logical because as technology investment expands, less land needs to be developed for agricultural use. In specifying the land development costs, a clear distinction should be made between the private and social costs involved with land development. Experience in the tropics has shown that the private cost involved with land clearance will be fairly small because one simply starts burning down the vegetation. So the main costs are social: halting the disappearance of the tropical forests is now viewed as for the common good. Consequently society becomes increasingly prepared to pay for their maintenance.

Precisely because the assessment of the land-development costs is based on its social shadow-costs, it is not an easy job to specify them. In the following we assumed that in any case forest protection requires investments, either for adoption of the forests, or for its maintenance and guardance, or to compensate the land owners or farmers for giving up their 'right' to penetrate the forests. All options boil down to a certain amount of money required per hectare of forest for its preservation. We assume that the investment one is prepared to make for preserving an additional hectare of forest will increase as the remaining forest area is getting smaller. In other words the social costs of land-development will generally increase with the total area developed for agricultural purposes.

Since preservation can be viewed as an investment (in maintaining the forest for future generations) the land-development cost curve will shift upward with an exogenous increase of the interest rate. A theoretical example may clarify this relationship. Suppose an individual farmer wants to enter the forest which is protected by environmentalists. He may succeed in convincing them to provide the desired forest area if he is able to provide for the resources enabling the environmentalists to acquire a piece of forest which is at least equivalent. If the interest rate increases the farmer obviously has to supply a larger amount of resources to 'buy' the same sized forest area for agricultural use. A similar reasoning applies for the price of concessions for commercial logging: if the wood prices increase, the price of the concessions will rise as well, ceteris paribus.

Both cost curves, the one involved with introducing agricultural technology and with land development, can now be combined into a total cost curve, to answer the question at which point total costs (related to a predetermined agricultural production level) reach their minimum level. This has been done in Figure 7.3. Obviously, the precise shape of the curve is a matter of empirics; its general form, however, is based on the above theoretical arguments. The figure shows that the total cost curve first decreases with an expansion of the agricultural technology investment per hectare because at first the productivity of those investments is high and at the same time the social costs of deforestation also; the latter because at a low level of agricultural technology a large area for agricultural production is needed, so that deforestation may to a large extent already have taken place. As agricultural technology moves forward, the return on agricultural productivity starts to decline as well as the social costs of deforestation – the latter due to the fact that the amount of deforestation becomes less.

If the farmer would take all the costs just specified into account, the size of his investment in agricultural technology at a given interest rate would be determined by the minimum of the total cost curve (point A in Figure 7.3). This optimum obviously will be affected by such factors as the interest rate and the price he has to pay for acquiring a unit of agricultural technology (for convenience' sake one could think of the application of fertilizer). A rise in the interest rate increases the land development cost curve and therefore shifts the minimum point upward to the right; a decline in the price of agricultural technology (e.g. through subsidized fertilizer) instead lowers the technology investment curve and therefore shifts the minimum point downward to the left. Both points are indicated A' and A" respectively in Figure 7.3.

Now the aforementioned iso-product curve and the total cost curve can be combined to translate the optimum for the individual farmers into the total area exploited for agricultural purposes on the one hand and the remaining forest on the other hand. On the basis of this model subsequent sensitivity analysis can be carried out for assessing the

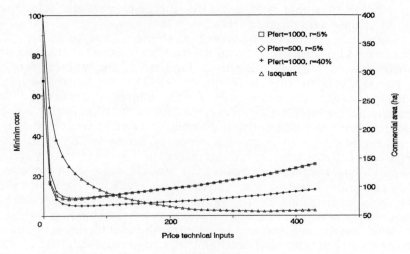

Figure 7.4a

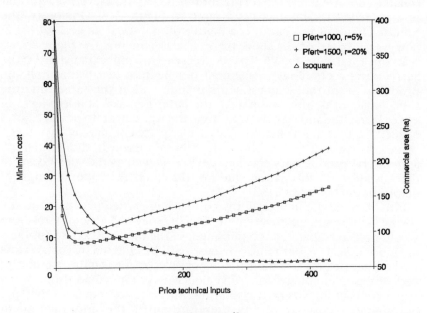

Figure 7.4b

impact of exogenous shifts in for instance the cost of agricultural technology, or in the interest rate. In Figure 7.4a the three total cost curves as represented in Figure 7.3 have been combined with an iso-product curve. The figure clearly shows that if the costs of agricultural technology decline (for instance through subsidies on fertilizer, or the transfer of agricultural technology through development projects), so will the area for agricultural production. The same holds for a rise in the (interest) costs of maintaining the forests. For in both cases the

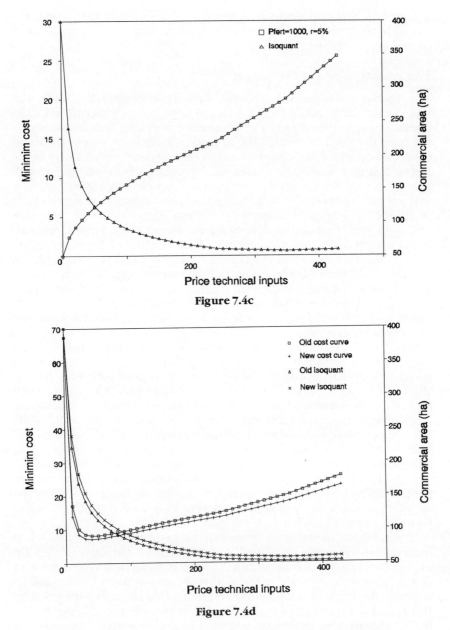

Figure 7.4c

Figure 7.4d

Figures 7.4a–d Combining the iso-product curve with cost curves

relevant points on the iso-product curve shift downward, indicated by the corresponding points A' and A" on the iso-product curve of Figure 7.3.

If both the cost of agricultural technology and interest rate move in the same direction, the impact on the total land area for agricultural production becomes unclear. Figure 7.4d illustrates the case where the size of both variables rises; clearly the final impact on land use depends on the aggregated vector.

The assumptions of the model

The above analysis is based on various assumptions. First, the assumption that the farmer has to pay a price for entering into the forest. The question arises to what extent this assumption is justified. If the costs involved are mainly social costs, the issue becomes how these costs can be translated into incentives for the farmers not to enter the forest. If such a 'translation' does not take place the farmer in fact will be only guided on the basis of his private costs (and benefits).

For if one assumes that in practice the costs of land development are small or even negligible from the individual farmers point of view, this obviously will have consequences for the shape of the aforementioned total cost curve. In the extreme case where land development costs are absent, the farmer's total cost curve will only be determined by the cost involved with acquiring agricultural technology. The cost curve could have a shape as indicated in Figure 7.4c (see also the earlier text). Any farmer who wishes to minimize costs will therefore keep expanding his land area; there will be virtually no limit to the deforestation process.

Secondly, the farmer will respond to new opportunities and improved incentives to increase land productivity. The equilibrium point may also alter due to a shift in the iso-quants following from a shift in the yield curve. If the level of knowledge about agricultural technologies increases, and if this additional insight is effectively transferred onto the farmers, the iso-product curve will shift downward as one can easily infer. Under normal circumstances this will lead to a shift of the equilibrium point whereby less area is required for agriculture. If the assumption underlying the iso-product curves, namely that the marginal rate of substitution declines with the level of technological inputs, holds true, the forest saving impact can be substantial, specifically if land development costs are relatively small. Figure 7.4d shows such a case.

It would be unrealistic, however, to assume that technological information can easily be passed onto the farmers, especially if they are engaged in subsistence agriculture. Social and physical characteristics of the agricultural sector often will represent a major constraint on reforms, the more so if land tenure relationships are highly inequitable. If agroclimatic conditions are highly differentiated, a readily applicable technology may be extremely scarce; a lack of resources can frustrate the transmission of agricultural technology to the subsistence sector and the administrative and managerial capacities can also fall short of enabling a successful introduction of new techniques. In short, almost any policy which aims to combat deforestation through increased land productivity at the forest fringes will require serious efforts to meet the various obstacles which can prevent a successful transmission of the information to the individual farmers. In addition one has to make sure that the technology, if made available, will actually be put into practice.

One important caveat should be made with respect to the above, however, namely that the assumption implicit in the reasoning that all land users have access to all means for land development does not hold in reality. Smallholders often cannot afford to buy fertilizer or pesticides, nor have they the most suitable technology for land clearance and infrastructure development at their disposal. The main reason is that due to their income level they cannot afford to invest too much because this would imply taking risks which might turn out to become unbearable. The same mechanism also reduces their chances of being able to lend on the local capital market if there is any, even if they would like to. Consequently the reasoning that physical land productivity can be equalized through varying degrees of land development is theoretical as far as the agricultural subsistence sector in developing countries is concerned.

In the model therefore, a distinction is made between the agricultural land area used for commercial agriculture, and the area used for subsistence agriculture. Precisely because the capital investment in land development in the subsistence agriculture is assumed to be rather modest, it seems fair to start from the assumption that agricultural productivity in the subsistence activities differs significantly from the productivity in the commercial agriculture. Note that the above distinction between the two categories of land use only relates to land productivity, and not necessarily to the scale of agricultural activities.

As was indicated earlier, there seem to be strong motives to treat subsistence agriculture as an economic activity which differs fundamentally from the pattern of commercial cropping described above. First, at the subsistence level the trade-off between the total land area developed on the one hand and technical inputs per hectare on the other hand, is less prominent because of the smallholder's lack of access to credits. Consequently, production is less advanced from a technical point of view. Moreover, the land development cost function as it is perceived at the subsistence level will differ from the one that is used as a starting point in the commercial sector. Specifically the use of capital goods, such as (heavy) machinery for clearance and the development of infrastructure will be severely restricted at the subsistence level. Land development at the subsistence level will therefore show much lower labour productivity and therefore higher labour costs.

The above implies that the behavioural pattern at the subsistence level demands a fundamentally different treatment in modelling. In our model the assumption has been made that the subsistence activities are in fact carried out by families, so the unit of production is very small. The main target of every unit is to try to produce such an amount of food that the likelihood of the actual harvest being insufficient to feed the family is very small. A second assumption is that agricultural productivity per unit will be the same across the units, but may differ between the regions. Finally, a third assumption is that the land area required to generate the desired food level per unit again is the same for every unit within a region. Consequently the land area needed for subsistent agriculture varies within every region in proportion to the size of the subsistence population.

SHIFTING AGRICULTURE: MODELLING THE COMMERCIAL VS THE SUBSISTENCE SECTOR

An important group of land users in tropical forest areas are those subsistence farmers who practise shifting agriculture. Any model of land use has to take the behaviour of these people into account. We have already developed models of the other two important categories of land users – timber producers and commercial agriculture – which are based on rational economic behaviour. Although the shifting agriculturists are certainly not irrational, their production processes are not governed by conventional market mechanisms. Furthermore, the fact that their production is continually shifting from one place to another complicates the analysis.

Our talks with experts in this field have enabled us to build up a preliminary picture of how these people obtain their food from the forest regions; here a very brief description is given. The rationale behind their behaviour is like that of any manager: to meet objectives while remaining subject to certain constraints. Without doubt, their first objective is to feed themselves; as a secondary goal one could perhaps add 'with the least effort'.

It can, therefore, be seen that population pressure will be the driving force of the whole subsistence sector. When population densities were low, such people lived in a sustainable manner in equilibrium with their environment. At the most primitive level they were hunter-gatherers, who really should be considered as part of the natural ecosystem rather than exploiters of it. However, shifting agriculture is also a sustainable system. A clearing is made in the forest and crops grown on it for a couple of years. As soil fertility drops, a new area of forest is cleared and the previously cultivated land left fallow to recover. What is of crucial importance is the length of the fallow period. On the one hand, after 30 years it is difficult for an untrained observer to tell that the land had ever been cultivated; furthermore, the soil fertility will have completely recovered. On the other hand, if it is returned to cultivation after only a few years, yields will be considerably lower.

Thus, the 'return-period' is of great importance, but one has to look more closely at the influences behind it. As indicated earlier, population increase is an important driving force which will inevitably demand more land. If not enough has lain fallow for the full period, then land will have to be brought back into use sooner than it ideally should. Is there anything that can mitigate this effect? One response is to move into 'virgin' forest that has not been previously used for shifting cultivation – provided, of course, that such forest is available and accessible.

Understanding the balance of activity between clearing new forest and returning to previously cultivated land is the major challenge of modelling shifting agriculture. As less and less virgin forest remains, the option of penetrating further into the forest disappears and shifting agriculturists will have to return to previously cultivated land, despite its low fertility. However, if there is still a substantial amount of forest to be exploited, other factors have to be considered. One of the most important is accessibility. How far do people have to travel to this new forest? How easy is it to travel? Are there any existing roads? Once the forest is reached, how easy is it to clear? If there are roads and the forest has already been partially cleared, the balance will certainly shift towards moving into new forest and against using previously cultivated land; these are precisely the conditions that exist in the train of commercial logging.

We will also have to look at the activities of commercial agriculture which will take land from other uses; and we will also have to consider the creation of 'waste land' as a result of certain land use policies (the complete loss of top soil through erosion and the spread of certain completely unproductive grasses).

In sum, we have only looked at very basic aspects of the problem. There are certainly further complicated interactions that could be of importance, for example agroforestry practices which combine arable farming and collection of naturally occurring forest products (fruits, rubber etc.). Even a relatively simple description of shifting agriculture presents a near insuperable challenge to the modeller. What can be seen is that, for a particular piece of land, time and history are vitally important. Every piece of land goes through a cycle of changing soil fertility. After initial clearance, fertility will be high. It will then drop rapidly during a relatively short period of cultivation. Subsequently, the land will be left fallow, during which period the fertility will gradually recover.

Description of land use changes

Many of the issues raised above can be represented diagrammatically, as shown in Figure 7.5. The upper diagram shows the breakdown of land use in the region of the humid tropics. It is, of course, purely symbolic and does not represent an accurate map of the areas used for different purposes, but it can be used to highlight some of the processes under discussion. The lower diagram shows the most important changes in land use that occur between one year and the next. There are, inevitably, some simplifications. Not all possible changes in land use are included: only cultivated land can turn to 'waste' land; forest is cleared only for immediate cultivation by shifting agriculturists; fallow land does not return to the forest. In addition, logging activities are not represented in detail; these are covered more comprehensively by the TROPFORM model and its derivatives. However, these diagrams give a good basis for discussion, and for model development, connected with our immediate concern of understanding and modelling agricultural land use in tropical forest areas.

Even this simplified diagram identifies ten different kinds of land use change. It can be seen, therefore, that there is a complicated system of

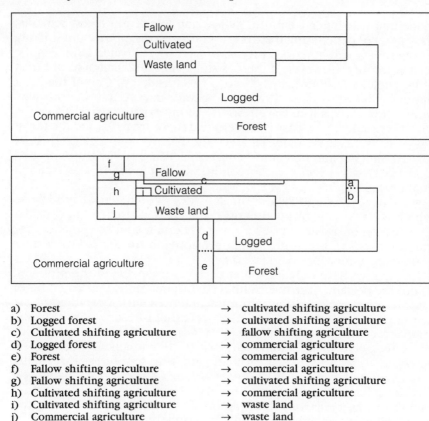

a) Forest → cultivated shifting agriculture
b) Logged forest → cultivated shifting agriculture
c) Cultivated shifting agriculture → fallow shifting agriculture
d) Logged forest → commercial agriculture
e) Forest → commercial agriculture
f) Fallow shifting agriculture → commercial agriculture
g) Fallow shifting agriculture → cultivated shifting agriculture
h) Cultivated shifting agriculture → commercial agriculture
i) Cultivated shifting agriculture → waste land
j) Commercial agriculture → waste land

Figure 7.5 Description of land use changes

interacting processes behind deforestation in the humid tropics. The challenge is to build a model that captures the essential mechanisms, and that gives insights into the workings of the system, and yet is relatively simple to develop and understand, and also does not require unobtainable data.

The modelling of land use changes

We shall now look at these changes of land use and examine how easy they will be to model. They are presented in another way in Figure 7.6. The letters designating different types of change are the same as in Figure 7.5.

The basic mechanisms for land use in commercial agriculture, based on the standard economic arguments of marginal analysis, have already been discussed earlier. The complication now apparent is that there are four different types of land that can be transferred to commercial agriculture: logged and unlogged forest, and the cultivated and fallow land of shifting agriculture. Questions are therefore raised concerning the marginal cost of new land to commercial agriculture and how the

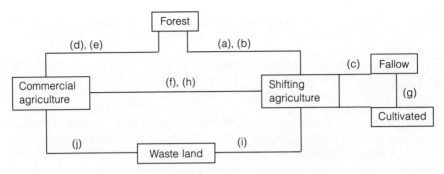

Figure 7.6 The modelling of land use changes

acquisition of this new land is divided among the four sources.

The politically contentious issue of land tenure is immediately raised by this question. Such an issue is almost impossible to model and is best treated as an exogenous component of a scenario; for example, what would be the effect of redistributing land from large landowners to peasant farmers? One does, however, have to make some assumptions.

Commercial agriculture will tend to use the cheapest available land first, but there are several factors that determine which land is cheapest. One important consideration is the location of land. Costs rise as distance from markets increases. Thus, to some extent, it is not important whether the land is currently cultivated, fallow or forest. Then, in addition, the actual preparation costs of the land have an influence. These costs depend on the current state of the land and on the crop to be planted. Clearly, virgin forest is more expensive to bring into cultivation than partially logged forest where roads have been built and the biggest trees already felled. It is probably also relatively easy to use the fallow land of subsistence farmers from their currently cultivated land (or perhaps employing them as waged labourers).

Bearing all these points in mind, initially exogenous, fixed proportions have been set for where new commercial agricultural land should come from. The marginal costs depend primarily on distance from markets and on local geography (e.g. mountainous terrain) but are also be influenced by some 'infrastructural variable' so that the improved access provided by logging activities lowers this cost.

We shall now look at the flows of land (a) and (b) from forest to the subsistence sector. The demand for land by subsistence farmers is primarily driven by population. As population grows, new land will be needed for cultivation. Some of this can be obtained by bringing more fallow land into cultivation (flow (g)). However, some cultivated land will have to be returned to fallow because of the fall in its fertility after a period of continuous cropping; this change in land use is represented by (c) in the diagrams. Therefore, any increase in food production is going to require new land not previously cultivated; this can only come from the forest.

Modelling this process raises the question of how much new land will be obtained from logged and unlogged forest, and how much fallow land can be returned to cultivation. Even though these subsistence agriculturists do not engage in market activities, it must be remembered that they are still economically rational. They will weigh up their choices and follow a course of action that will give them the greatest benefit for effort expended. If there is little forest left, or if it is a long distance away, fallow land will have to be used, even when the fallow period is much shorter than ideal. If the forest is readily accessible, then these farmers will extend their farms into the forest. Furthermore, if logging activities have cleared roads and removed many trees already, then the incentive to expand into the forest is even greater.

This choice can be modelled in terms of conventional economic analysis. A subsistence farmer must first feed his family, thus a certain amount of food has to be produced. He has a choice of using new forest land or recultivating fallow land. The balance of this choice will depend on the relative costs. However, because he is not operating in markets for either his produce or for factors of production (e.g. labour), 'costs' to him can only be measured in terms of his and his family's time and effort. If the fallow period has been too short, yields will be low and he and his family will have to work harder, tending a larger area of land to produce the food needed. Similarly, if he has to travel a long distance to the forest, and perhaps compete with others when he gets there, the balance of costs will swing towards recultivating fallow land.

Although one individual family will most probably either go to the forest or use fallow land, the aggregate behaviour of the whole subsistence sector can be demonstrated using a standard isoquant curve. This is shown in Figure 7.7 where the curve shows all the production possibilities for growing the amount of food required by all in the sector. The total perceived 'costs' are represented by an isocost

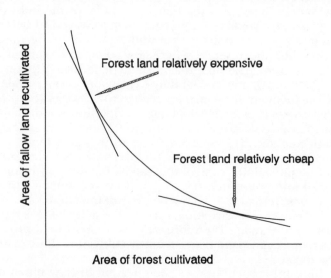

Figure 7.7 Subsistence sector production possibilities

straight line which, at the optimum, must be tangential to the production curve. Thus the balance between going into the forest or using fallow land depends on the relative costs which determine the gradient of the isocost line.

At first sight, the development of these production and cost functions does pose problems of data collection. However, it is possible to obtain figures for current yields in subsistence agriculture. Furthermore, there is information on how many years in succession a particular place of land is cropped, and on the ideal fallow period. The ratio of fallow land to cultivated land gives an indication of 'return period', the fallow interval between cultivation periods. For example, if at equilibrium a piece of land is used for three years and then left fallow for thirty, the area lying fallow will be ten times as great as the area under cultivation. Using such a measure it will not be too difficult to interpolate the current obtainable yield and thus the 'costs' of using fallow land.

The cost of expanding into forest areas will have two main components, travel to the forest and clearing the forest for cultivation. These, in turn, will be determined by a geographical factor (area of forest remaining, distance to the forest, local topography) and some measure of infrastructural accessibility (roads, partial logging). If we make some assumptions about the shape of the area used by subsistence farmers (e.g. it is a rectangular strip alongside the edge of the forest), it is easy to infer the distance to the forest for the average farmer. The state of the infrastructure will influence both the travelling time and the ease of clearing the forest once there. One can make some reasonable estimates of all the times involved but, because one is dealing with a complicated and conflicting set of human attitudes, it will be necessary to introduce one exogenous multiplier to represent the different attitudes to effort spent clearing the land and effort spent travelling. This can then become a potentially important input to different scenarios.

Finally, we come to the degeneration of agricultural land into waste land. This is relatively easy to model, or at least to define parameters for. One can assume that a certain proportion is lost each year; for example, it seems apparent that cattle ranching in cleared forest areas cannot continue for more than seven years because of soil erosion. Therefore an annual loss in the range 10–20 per cent would model such a process quite well. It will be necessary to keep track of how much former forest land is being used for commercial agriculture. Furthermore, assumptions will have to be made about the uses to which this land is put as not all will have such damaging effects as grazing livestock. Tree-based agriculture (e.g. palm oil) is much more in harmony with the local environment and is thus sustainable for far longer periods. Arable crops, such as corn or cassava, will cause greater damage to the soil which, however, may be in part recoverable by the use of artificial fertilizers.

The loss of land due to shifting agricultural practices, for example to non-productive grasses, needs further investigation. However, the same approach to parametization of assuming a constant proportion lost each year would be adequate. Since we do have a measure of fallow period, this could also be allowed to influence the rate of loss.

Conclusions

We have described the most important modes of land use in humid tropical forest areas and investigated the necessary steps that must be taken to model such changes. The task is not easy, but neither is it insuperable. As much data as possible have been collected on the costs of forest clearance both for commercial agriculture (in terms of money) and subsistence agriculture (in terms of time). In addition information has been gathered about what sort of yields are obtainable in shifting agriculture and how they are reduced by lowering the fallow period. Finally, data have been collected, and assumptions made, about the rate of loss of land which becomes unproductive waste.

In such an area of study it would not be possible to obtain all the data needed precisely. However, what we will be able to achieve is a mathematical description that depends on a handful of parameters. Furthermore, in consultation with experts, we will be able to say within what ranges these parameters must lie. As a result, the model will be able to highlight which mechanisms are crucial to deforestation. Finally it will point out which of our assumptions and parameters are most critical and thus where further research must be aimed if we are to save one of our planet's great natural resources.

In Figure 7.8 the main mechanisms underlying the use of land for commercial and subsistence agriculture are presented.

MODELLING AND SIMULATION OF SOCIO-ECONOMIC PROCESSES: SARUM

SARUM is a global economic model developed in the mid-1970s by the Systems Analysis Research Unit (SARU) of the UK Department of Environment and Transport. The model is a systems dynamic model inspired by the Limits to Growth (Meadows, 1972) discussion, but containing important economic feedback mechanisms. Since the model was completed, it was reviewed in the IIASA framework and adopted for a variety of applications in a number of projects, including:

- the OECD Interfutures Project (OECD, 1979);
- the Food-Energy Global Modelling Project at the UN University;
- the Australian Resources and Environmental Assessment (AREA) Project;
- the Global Models and the Policy Process (G-MAPP) Project at the University of Hawaii; and
- the future financial requirements of the less and of the least developed countries projects carried out by the University of Groningen (Jepma, 1990 and 1992).

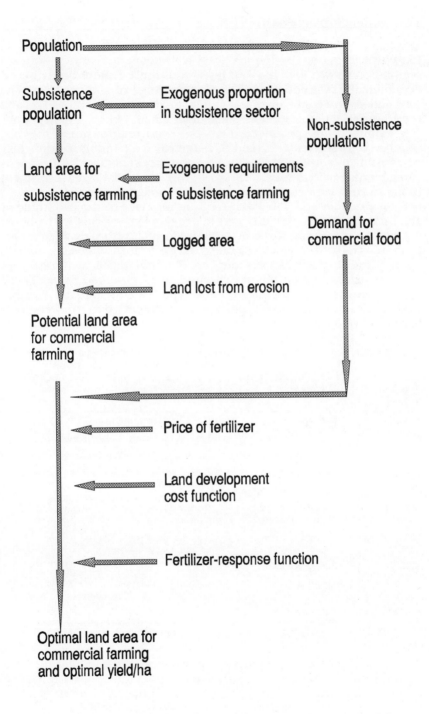

Figure 7.8 The mechanisms underlying land-use in the commercial and the subsistence agricultural sectors

The model characteristics

The model, which has strong neo-classical characteristics, represents the world economy by ten geographical, economic and/or political regions, as shown in Table 7.1. In each of these regions a number of economically distinct sectors produce a range of commodities. The production process involves the employment of labour and capital and the consumption of resources, some of which may be in limited supply. The commodities produced may be exported to other regions, purchased by sectors as capital or as intermediate inputs, or they may be consumed by the population. The population obtains its income as capital returns and as wages, and its savings are available for investment in the sector's capital equipment, and for foreign payments.

Economic interaction between regions may take the form of trade or aid. Demand for a commodity within a region may be satisfied by home producers or from imports, depending on relative prices. Distortions from free trade from whatever cause (tariffs, transport costs, political, cultural and language barriers, etc.) are incorporated using the concept of trade biases, which modify the perceived purchase price. These biases may be regarded as important policy instruments. Flexible exchange rates ensure that regions are not permitted to run persistent balance-of-trade imbalances.

Table 7.1 Regional division of countries in SARUM; most important countries per region*

Notation used in SARUM	Region	Most important countries
NORAM	North America	United States and Canada
EUR	Western Europe	France, Germany, United Kingdom
DEVPA	Japan and Australasia	Japan, Australia, New Zealand
LATROP	Tropical Latin America	Bolivia, Brazil, Colombia, Peru, Venezuela
LAREST	Rest of Latin America	
EURCPL	Formerly centrally planned Europe	(former) Soviet Union, Hungary, Poland, (former) Yugoslavia
AFTROP	Tropical Africa	Angola, Cameroon, Central Afr. Rep., Congo, Gabon, Tanzania, Zaire
AFREST	Rest of Africa	
ASTROP	Tropical Asia	Indonesia, Malaysia, Thailand, Myanmar (Burma), Papua New Guinea
ASREST	Rest of Asia	

* A detailed country classification will be presented in Appendix 5.

Sectors are the basic production units within the model and are aggregated from industries whose requirements for labour, capital and inputs are expected to develop along similar lines. The current version of the model has four sectors, which are shown in Table 7.2.

Sectors produce their output commodities by employing labour and capital and consuming certain inputs. Both capital and the consumable

Table 7.2 Sectoral division of countries in SARUM

Notation used in SARUM	Sector
FOOD	Food
NATPRD	Natural products
INDUS	Industry
SERVIC	Services

input commodities are obtained as the outputs from other sectors. Input requirements are specified as coefficients derived from input/output tables and capital requirements are specified as distribution of the various capital commodities. A Cobb-Douglas production function is assumed and hence the optimum mix of labour and capital is determined as a function of their prices (wages and interest rate).

Sectors grow and decline in response to three effects. First, demand patterns may change in response to rising income or changing tastes. Secondly, relative production costs may change as a result of resource depletion, environmental constraints or the emergence of new production processes. Thirdly, increased incomes may raise labour costs and favour those sectors most able to substitute capital for labour.

It is possible to model various different types of sectors:

- resource limited sectors are those in which the cost of producing a unit of output increases with the total cumulative production;
- flow limited sectors are those in which the costs increase as a function of the realized rate of production.

The model assumes that there is a competitive market for each commodity, but supply and demand need not always be in equilibrium. A system of stocks serves as a buffer between production and consumption. Prices are inversely related to stocks in such a way that, when stocks are at their desired levels, prices are proportional to marginal costs. If stocks are above the desired level prices fall, and vice versa. Prices act as signals to producers and consumers to regulate supply and demand, thus tending to return stocks to their desired levels.

The output of a sector is given by a Cobb-Douglas production function of capital and labour. Managers are assumed to be profit maximizers and therefore attempt to vary labour and capital in such a way as to pay these factors their marginal product. If the marginal product of labour is greater than the wage rate the labour force is increased; if less, it is reduced. The amount of capital is not directly under the managers' control and varies at a rate determined by investment and depreciation. Inputs to production other than labour and capital are supplied in fixed proportions using data from input/output tables.

The model allows a number of aspects of technological change to be simulated. The technical characteristics of existing industrial sectors are embodied in the Cobb-Douglas production functions (which describe labour and capital productivity) and in the input–output coefficients (which specify the sectors' input requirements). Both may be varied to reflect anticipated changes.

The advent and growth of radically new technologies can also be modelled. This is done by defining 'seed' sectors with appropriate technical characteristics. These sectors may not, initially, be viable and may have to be subsidized until such time as changing conditions make them competitive.

Regional income, or gross regional product, is calculated as the sum of values added by all sectors. This sum, modified by any net foreign aid receipts, then becomes the regional expenditure and is allocated to investment, industrial subsidies and final consumption. There are no financial institutions or stocks of money in the model so that investment is determined for each region as a function of per capita income, and is distributed among the sectors on the basis of their marginal profitability in such a way that sectors with the same marginal profitability grow at the same rate.

Subsidies are provided to sectors for two purposes: to retain a particular industrial capability in the face of foreign import competition; and to support the development of a new industry during its early stages. In both cases the subsidy is paid if output falls (or remains) below a user specified level. Subsidies take the form of free gifts of capital to replace capital lost via depreciation.

Foreign aid may be given for three purposes:

1) General aid, 'with no strings attached', is added directly to the recipient region's total expenditure and distributed according to the usual rules.
2) Investment aid is added to whatever sum the region allocates to investment out of its total expenditure. Total investment, including aid is then distributed among the sectors.
3) Aid may be earmarked for investment in a particular sector, and is added to whatever sum the region allocates to that sector out of its total investment budget.

The population of the region provides the labour force for the sectors. As already noted, sector managers attempt to maximize profits, in part by adjusting their labour force so that the marginal product of labour equals the sector wage rate. As the demand for labour increases, wages rise according to an empirically based relationship. This has two effects. First, the higher wages restrain the increasing demand for labour through their effect on production costs and product prices. Secondly, labour is attracted from other sectors with lower wages. These effects are governed by delays associated with recruitment and wastage and particularly with retraining for works in new sectors.

Demand for a commodity within a region is satisfied by local production or from imports, depending on the relative prices and quantities produced in each region. Deviations from the basic model, representing such effects as tariffs, transport costs and preferential trading agreements, are incorporated using a set of trade biases which modify the trade flow between each pair of regions. These trade biases may be varied and constitute important policy instruments for the study of alternative trading patterns and economic groupings.

Regions are not permitted to run persistent balance-of-trade deficits. The main deficit control mechanism acts via the use of exchange rates which vary so that the exports of deficit regions become cheaper while their imports become more expensive.

A large number of behavioural relations derived from standard economic theory form the basis of SARUM, of which only a few are described above. SARUM offers a wide range of instrumental or policy variables such as:

- population growth rates;
- the rates of technical progress and investment;
- consumption;
- trade policy;
- investment policies;
- aid policies, in terms of aid volume and its sectoral and regional allocation; and
- policies affecting income distribution.

The user of the model can change the values of these policy variables as well as make different hypotheses of what will happen in future economic conditions. Due to its long history, frequent application and continuous updating, the model has been extensively documented at various occasions. For a detailed description of the model structure, characteristics, validation, database, etc., the reader can therefore be referred to: SARUM (System Analysis Research Unit, 1978); Gigengack, Jepma et al. (1985); and Poldy et al. (1986).

MODELLING MIGRATION[1]

Land-use modelling plays an important part in the integration of SARUM and TROPFORM into IDIOM. In modelling land use, one needs to know where people and economic activity are located, however. The question arises for instance how many people are engaged in agriculture, and how many of them can be classified as part of the subsistence sector. This leads us to the topic of migration.

The issue of rural–urban migration holds an important place in the literature in the field of development economics. Some main aspects of this literature and its relevance to land-use modelling will therefore be reviewed. We have already discussed past trends of urbanization and past and current policies directed towards migration. We will now outline the various discussions that took place during our search for a suitable structure to model the migration process.

Modelling migration: background considerations

The primary reason why including land-use elements into our model was imperative for a better understanding of the process of

[1] See also Parker, 1991, 1992.

deforestation is that this information is vital for understanding the causes of the current loss of tropical rainforests. Although TROPFORM provides detailed estimates of losses due to timber extractions, the model deals with the competing uses of forest areas in a very rudimentary, and essentially exogenous, way. The impetus behind our approach is that deforestation is basically determined by human behaviour through a complex interaction of socio-economic factors of which commercial logging is only one, and perhaps a minor contributor.

Forest can only be felled by people who are actually there. Apart from the commercial loggers, who are the others who make their living in the forest? Industrial activity is unlikely to take place in such areas (being away from commercial centres) unless there is some important resource located there. The manufacture of timber and timber products is one obvious possibility; the only other one likely to be of any significance is the extraction of minerals. Apart from these, the most important resource is the land itself that can be used for agriculture. Therefore, any study of forest land use must look at agriculture. There are, of course, many types of agriculture and any of them can, potentially, use land that has been cleared of forest. Broadly speaking, agriculture can be classified in terms of what the land is used for, and the economic arrangements for the use and distribution of the products. In our modelling it is not proposed to look in any finer detail than a classification based on arable/livestock and commercial/ subsistence. Indeed, in the first instance, the arable and livestock operations will be combined.

Commercial and subsistence agriculture cannot, in practice, be completely separated; many people will both produce some food for their own consumption and work for commercial farmers, especially on a seasonal basis. Also, many subsistence farmers will sell surplus produce and buy simple equipment. However, there are differences in technology of production and the sources of demand, and thus it is a useful distinction to make. Much of the demand for commercially produced food will come from the urban population; therefore the land area needed will depend on the degree of urbanization and the incomes of the customers. The development of subsistence agriculture will depend on the size of the population in this sector, and that will be critically influenced by migration. In summary:

- people need food;
- the production of food needs land;
- the yield per hectare, and thus the land area needed, depends on the agricultural system (commercial, subsistence);
- the location of the land use depends on where people are;
- therefore any deforestation due to food production depends on where people live and what form of agriculture supplies their food needs.

There has been a steady flow of work related to migration for some decades, covering both developed and less-developed countries, and also within-country and between-country migration. For our purposes we are primarily interested in migration between rural and urban areas within a country. Many of the principles of migration modelling are independent of the state of development (e.g. wage differentials and the analysis of labour markets). However, tropical forests lie almost exclusively within the developing and the newly industrialized countries and therefore our main focus must be on the movement of people between urban and rural areas in these countries.

Traditional models of rural-urban migration

Lewis' approach
Rural-urban migration is closely linked with the labour-markets in the urban (commercial) and rural (subsistence) areas. In a model postulated by Lewis (1954), the labour market is being modelled in the following way: different labour-markets for the subsistence and the commercial sector.

The Lewis model starts with the assumption of a dual economy with a modern exchange sector and an indigenous subsistence sector, and assumes that there are unlimited supplies of labour in the subsistence sector in the sense that the supply of labour exceeds the demand for labour at the subsistence wage; that is, the marginal product of workers in the subsistence sector is equal to or below the subsistence or international wage so that a decrease in the amount of workers would not lower the average (subsistence) product of labour and might even raise it.

It has even been argued that the marginal productivity of labour may be zero or negative in an economy still in a fairly low stage of development, experiencing an rapid growth of population. One of the features of agriculture is that it is an activity subject to diminishing returns owing to the fixity of the supply of land. If there is a rapid population growth and labour has little employment opportunity other than on the land, a stage may be reached where the land cannot give further workers a living unless existing workers reduce their own hours of work drastically.

There are three main escapes from the tendency towards diminishing returns and zero marginal productivity in agriculture. First, by productivity increasing faster than population through the absorption of more and more of the agricultural population into industry. Secondly, by technical progress in the agricultural sector, increasing labour's marginal product. Thirdly, by capital accumulation which can raise productivity directly and which can also be the source for technical progress.

The underlying assumption in Lewis' theory is that employment growth in the capitalistic sector is proportional with capital formation. However, if the profits made in the capitalistic sector are invested in labour-saving technologies, this will not hold. If, on the other hand, the above reasoning does hold, the growth of employment in the capitalistic sector will, in the end, result in a competition between the subsistence sector and the capitalistic sector for labour; this will cause the agricultural sector to become more and more capitalistic.

In Lewis' model, the wage differential between the commercial and the subsistence sector serves to attract labour to the commercial sector. Migration from the subsistence sector to the commercial sector only takes place when a worker in the subsistence sector is able to get a job in the commercial sector.

What we have seen in the last few decades is that the urban-rural wage differential has widened considerably beyond 30 per cent (Thirlwall, 1983). 'In the same period there has been rural-urban migration at an unprecedented scale, but the expansion of the industrial sector has not generated sufficient employment. Migration has thus served to transfer unemployment from rural to urban areas.' (ibid.)

Developing countries' governments' policies have implicitly led to rural–urban migration, due to the preferential treatment of cities for industrial development, infrastructure investment, pricing policies, social services and for food and other subsidies.

Due to the lack of sufficient manpower, attempts to develop the rural areas failed. Therefore, the rise in agricultural productivity could not prevent people migrating to urban areas. Several attempts have been made to restrict rural–urban migration by setting a minimum rural wage level in order to reduce the rural–urban wage differences. The result of this policy measure was an accelerated farm mechanization, however, which led to rural unemployment and therefore a more rapid rural–urban migration.

Todaro's approach

A migration model developed by Todaro is seminal to much of what has followed since (Todaro, 1969, pp.138–148; Harris and Todaro, 1970, pp.126–142). The model was developed in order to explain a phenomenon that is part of the received wisdom of migration studies in Third World countries: rural-to-urban migration continues at high levels even when urban unemployment remains high or, indeed, is still increasing. Todaro's basic thesis is that, despite all the social and psychological influences, the decision whether or not to migrate is primarily an economic decision. In contrast to the Lewis two-sector model, which assumes that there is an unlimited supply of rural labour willing to move to the cities, the Todaro model postulates that potential migrants look at both the wage differential and the probability of getting a job when they reach the city. In the general description of the model, it is stated that this decision will be influenced by longer term considerations: the migrant will, in general, be prepared to wait a certain length of time before getting an urban job.

Obviously, the longer the expected wait, the less attractive it is to migrate. This balance is explicitly modelled by calculating the expected difference in wages, where 'expected' has its statistical meaning of an average value (value multiplied by probability).

Essentially, potential migrants look at urban wages, multiply them by the probability of getting a job, and compare the resultant figure with their rural wage. A final link in the Todaro model is provided by a relationship between urban unemployment and the probability of a newly-arrived migrant finding a job. In the mathematical analysis it is assumed that this probability is inversely proportional to unemployment rate (the excess supply of labour as a fraction of total employment).

A study of Kenya by Ian Livingstone found that the rural areas served as a very effective 'sponge', absorbing and retaining a great part of the additional population and labour. Diminishing opportunities for agricultural expansion at the extensive margin resulted in a declining average amount of land held per person for the last decades. This study clearly shows that farm size affects the efficiency of production. Hectares per worker varied from under 0.5 up to 8.0 hectares and over for smallholders. According to Livingstone, 1990, the average household per member in the lowest land category was 60 per cent of that of the largest land category.

The structure of the Todaro model seems simple – and the relationships within it at first sight not too unrealistic – but we do need to ask whether it can reproduce the behaviour observed in the real world. Todaro's main concern is to show that an increase in the provision of urban jobs can lead to greater urban unemployment. The mathematical analysis shows that this certainly can happen, but it is also useful to understand the process conceptually. The inferences drawn from the equations can be explained as follows:

- Additional job creation in the cities increases the probability of a migrant getting a job.
- The increase in probability increases the expected wages perceived by the migrant.
- The increase in expected urban wages increases the rate of migration.
- The increase in migration, for most realistic combinations of parameters, exceeds the number of new jobs created. Hence urban unemployment increases.

The model has been developed in various ways made by subsequent researchers. Much has been made of the concepts of 'formal' and 'informal' sectors, particularly as a categorization of the urban labour market. One paper which extends this idea to six-fold classification is by Santiago and Thorbecke (1988, pp.127–148). They introduce an additional sector, rural non-farm, which is itself further divided into formal and informal. They then present comprehensive data on employment and earnings in these sectors at four points in time (1950, 1960, 1970 and 1980). These show very clearly the progress of urbanization and the growth of the formal economy in Puerto Rico. Their work, however, does not show the causes of such changes.

According to Todaro's thesis, differential wage rates should play an important part in determining migration and, indeed, wage-rate differences are apparent in some of the tables presented in Santiago and Thorbecke, 1988. According to these authors the ratio of urban to rural wages has fallen steadily (from 2.51 in 1950 to 1.23 in 1980 in Puerto Rico). On the other hand, the ratio of formal to informal wages has risen from 1.80 to 4.34 over the same period. Furthermore, analysis of variance shows that a two-sector, rural-urban categorization has never explained much of the variation in earnings; at any time in the whole thirty-year period, the proportion of the variance explained by within-sector variance has never fallen below 93 per cent. Indeed, the proportion explained by the variance between the urban and rural sectors has stayed very steady at 6–7 per cent. The between-sector differences for a formal-informal two-sector classification have been much higher, and rising (14.5 per cent of the variance explained in 1950 and 62.7 per cent in 1980). One has to be a little cautious when interpreting such figures as the definitions of formal and informal are inevitably somewhat arbitrary; in this particular study it was based on the number of employees in each establishment. This problem of classification is considered below.

A continuing approach to migration studies, with its basis in the human-capital model, is the use of regression analysis in which the propensity to migrate is explained in terms of demographic and 'life-experience' variables. Thus the effects of age, number of children, education etc. on migration can be deduced. Examples of such studies appear quite numerous (Taylor, 1986, pp.147–171, Shields and Shields, 1989, pp.73–88, Pessino, 1991). The general impression is that these models indicate broadly who will move, but do not say a great deal about the underlying causes which make migration an option worth considering. Such variables as age and number of children seem important, but the evidence on the effect of wages is ambiguous. For example, Taylor states that 'There is evidence that higher household incomes discourage internal migration' whereas Shields and Shields conclude 'that families with higher income are more mobile.'

Kannappan's view
It is clear that migration is a complex issue with many determinants. In a paper written by Kannappan (1985, pp.699–730), the central thesis is that much of the analysis of migration has been constrained and moulded by prior assumptions. Views on the importance of dualism (formal/informal), and assumptions about urban unemployment, have prejudiced the theories and the research. If you assume that there is a clear divide between formal and informal urban economic activities, and then force your data into this categorization, often in a rather arbitrary way, you will very likely be able to draw some conclusions about the differences between the two sectors. However, these conclusions depend on the initial assumption that there are two distinct sectors. With detailed reference to many studies, Kannappan demonstrates that these sectors, which are often assumed to be dichotomous, are in fact very closely intermingled. There is a wide spectrum of wage rates in all sectors and it is likely that someone in the informal sector can earn more than someone employed in the formal sector. And there is certainly no evidence, as some writers have

proposed, that there is a two-tier queuing system in which people wait while unemployed, then move to the informal sector where they stay until a formal sector job is available.

Analyses of rural-urban wage differences also show no clear, large differences, especially when the wide range of wages is taken into account. This is corroborated by Santiago and Thorbecke's analysis of variance studies mentioned above. One of the central tenets of 'conventional' migration theory is that these wage differentials exist and that they cause excessive migration, and thus concomitant urban unemployment. By reference to various studies, Kannappan challenges this assumption that excessively high urban unemployment universally exists in Third World cities. Support for this view can be found in an article by Shukla and Stark (1986, pp.139–146) which, by means of a simple profit-maximizing model, shows that urban employment could be too low. They then analyse the best policy instruments for increasing urban employment.

What, therefore, are the conclusions that one can draw about migration and the operation of labour markets? Kannappan concludes rather convincingly that 'migration is a rational but diverse process of studied responses to changing economic conditions and requirements.' There is a much better network of informal information flows than has often been assumed: 'word-of-mouth contacts, friends, and relatives . . . are the most common means by which workers find jobs; [the informal channel's] role is more pervasive, perhaps indispensable, in developing nations, given the greater strength of traditional ties and variable attributes among workers and jobs'. In the end, this view on urban labour markets can be condensed into the following three points.

1) Prevailing sectoral aggregations fail as examples of dual labour markets.
2) Urban wage rigidity is not typical, and there are enough indications of a dispersed wage structure tending towards equilibrium.
3) The diverse migratory and employment channels and probabilities also suggest decentralized and flexible adjustments rather than a chronically unresponsive unemployment rate.

These leave us without an economy-wide theory of urban unemployment, however, not very encouraging for a potential modeller!

Implications of the literature

What implications does the above have for our land use modelling? The problem is that the literature seems to be pointing in several different directions at once. The urban labour market does appear very heterogeneous and thus any classification system is very likely to be unrealistic. One can infer from one part of the literature that the labour markets are more efficient than might be thought and that rural-to-urban migration is a rational response to economic development. The next step is to infer that such migration is demand-driven. It is the growth of the urban economy that places a demand on the labour market. Hence, if urban job creation is large enough, people will be drawn from the countryside into the cities. Such a process will have considerable ramifications. The growth of the urban population will require an increase in the commercial agricultural production. There could then be a consequential increase in the labour force of commercial agriculture (or, at least, not such a large technology-induced fall). Given the increase in urban population, there will thus have to be an outmigration from rural areas, especially from the subsistence-farming sector.

Such a demand-driven view of the operations of the labour market can only be part of the story; one still has to answer the question, why do people migrate? People do have information about urban conditions, albeit imperfect. Many migrants retain links with their place of origin and, which is not always given as much attention as it deserves, a large number of people make more than one migratory journey. It is interesting to note the point made by Pessino in her paper (Pessino, 1991) that cities with the highest in-migration also have the highest out-migration (and this is true in proportional terms, not just absolute). Many migrants will find that the streets are not 'paved with gold' and will return home. Others will try a succession of locations to live; indeed, this can be a profitable way for individuals to build up capital (Sewastynowicz, 1986, pp.731–753). These migrants will, of course, bring information with them about their experiences and thus will contribute to the decisions of other potential migrants.

Even given the many reservations that have been presented in the literature, one cannot completely ignore wage differentials as one important driving force of migration. Indeed, Kannappan would not deny the rationality of pursuing economic benefits in the cities; it is just that you cannot measure the urban job market in a simple way. Potential migrants will look at their own situations and abilities, and compare them with what they have heard about urban conditions. They may well have family connections or particular skills that make it worthwhile moving even when, at the destination, average wages are low or unemployment is high. If particular urban industries are growing, they will need additional labour, and will thus need ways of attracting extra workers. Several examples can be given of how this is done. The employers will, for example, use contacts with their existing employees, or use a job broker. Essentially, however, if they want to attract people from other areas, they will have to give them a financial incentive. However, because of the heterogeneous nature of employment conditions, it is quite possible for them to offer a sufficiently high wage that still lies within the existing urban spectrum, or to find people who have below-average standards of living in the source areas for labour. Such an argument explains how migration can be induced without there being a clearly defined differential in average earnings.

Models that include return migration could be very useful for the modelling of land use in tropical-forest areas. There does appear to be prima facie evidence that agricultural development does occur in tropical forests in the wake of commercial logging activity. It is likely, therefore, that the informal information networks would communicate this information to the cities. Since out-migration from cities is high, some of the migrants would leave the city for the opened-up frontier regions. In essence, the process is no different from that of rural-urban migration: if there are job opportunities, people will find them.

Migration modelling and tropical forests

We now have to ask the question, what form of migration modelling is most appropriate to our modelling efforts? What we are looking for is a way to model the number of people in subsistence farming (and thus, indirectly, the land area used), and also the area required by commercial agriculture.

Let us first investigate the implications of accepting Kannappan's view that labour markets are not completely inefficient and that workers will move to where there are job opportunities. SARUM does generate labour force figures for every sector in every region, and these can be considered as a measure of the demand for jobs, or of the provision of job opportunities. As economic development proceeds, the labour forces in manufacturing and services will increase, implying a larger urban population. Consequently, a measure of the urban population is implicitly present in SARUM. And, as this measure changes over time, it would, in theory, be possible to deduce figures for net rural-urban migration.

The present SARUM data set excludes subsistence agriculture; production in the food sector is purely commercial. Since the population figures in the model include the whole population, subsistence farmers are somehow 'hidden'; they will, in fact, have been distributed around all the other sectors. It might, therefore, be more realistic to initialize the data in SARUM using population figures that exclude those in subsistence farming. However, it would then be necessary to modify the labour-market mechanisms within the model once the simulation started. But you could not cut off the subsistence farmers from the general labour pool forever; it is an uncontested fact that such people do migrate and join the rest of the economy.

Another problem is that our land-use model will work at the level of TROPFORM regions which, where they contain tropical forests, are individual countries. Any figures derived from SARUM will pertain to much larger regions (e.g. less developed sub-Saharan Africa). It would be best to have a separate labour model of every TROPFORM region. This would divide the workforce and population (which includes dependants) into subsistence agriculture, commercial agriculture, and 'other commercial activities.' It may also be necessary to further divide the 'other commercial activities' sector into its rural and urban components. These sectors of the labour market would provide job opportunities which determined how many people worked in each. The rate of change of job opportunities in each labour sector was determined by SARUM. That is, instead of trying to apportion gross SARUM figures over individual countries, one assumes that if the

African labour force in commercial agriculture was rising at 1 per cent per annum, then the commercial agricultural labour force in Zaire would also rise at 1 per cent per annum. It would not, in fact, be too complicated an extension to include 'elasticities' which determine the ratio of growth rates in the small TROPFORM regions to that in the encompassing SARUM region.

How could we model the critical figure of population in the subsistence agriculture sector? Our argument here, based on Kannappan's views, is that labour follows job opportunities. As suggested above, the labour models for each TROPFORM region will generate figures for job opportunities in non-subsistence sectors. It is, therefore, possible to model out-migration from the subsistence sector, essentially by treating subsistence employment as a residual. There would, however, have to be some form of control to ensure that negative quantities did not result. The labour model would necessarily contain a simple demographic element and thus there would be population-growth pressures within each sector. Consequently, if the growth of population in the subsistence sector outstripped the job creation in the alternative sectors, there would be greater pressure on land use by subsistence farmers. This would be reflected in greater development of new land (by forest clearance, for example), or by reduced fallow periods.

The simple labour model proposed here looks reasonable as a way of modelling net rural-urban migration. However, it does not explicitly incorporate return migration, and certainly not multi-stage sequential migration. We do want to change the balance of migration as forest areas become more accessible, however. The job-opportunities framework would seem a reasonable basis for modelling such an effect: the provision of infrastructure for subsistence farming creates job opportunities in that sector just as much as capital investment creates industrial employment. What we need to do, therefore, is develop a model of inter-sectoral labour flows that is driven by job opportunities, numbers already working in the sector, and some measure of sectoral unemployment. The last of these would provide the main control mechanism in that it would try to equalize the number of dependants per job in each sector. The mathematical formulation to be developed would almost certainly bear a strong resemblance to the current employment model within SARUM, with its 'fractions trained' and inter-sector mobility.

The question does arise, how do we model job opportunities in the subsistence sector? In some ways, this is simple: if a person returns to the subsistence sector, he or she has immediately created a job – that of feeding themselves! However, we have to think more precisely about what is necessary to provide a job. A definition that covers all types of job, whether on the formal economic system or not, is that a 'job' is the coming together of various factors to produce a good or service that somebody wants (and is willing to pay for, though not necessarily in cash). Labour is obviously always an important factor of production but there will, almost always, be an element of capital as well. Employment in subsistence agriculture certainly needs simple tools but, above all else, it needs land.

The supply of job opportunities in subsistence agriculture is thus primarily dependent on the amount of land available. Land availability will have to take into account how easily available that land is; it is not easy to farm a piece of land 1000 kilometres into the Amazonian rainforest! Assuming the necessary information was available, a good measure would be the amount of land that a person could obtain for food production within a limited time period. Land that could be made ready within a few weeks would probably provide a viable job opportunity; if it took more than a year, then there would really be no realistic job. This approach does work in one of the most interesting situations, that when logging activity clears the way for subsistence farming. It is possible to make land ready for farming in a much shorter time if someone has already built the roads and cleared most of the trees. Consequently, more subsistence farming jobs are created and thus, in our job opportunities model, migration to the subsistence farming within forest areas is increased.

The Todaro wage-differential approach

In this model, migration is driven by expected wage differentials. In order to be operationalized it needs sectoral wage rates and sectoral unemployment rates as inputs. Furthermore, the functional form relating migration rate to expected wage differential has to be specified. Assuming this is possible, we do have to consider whether the model can be made to reproduce return migration as opportunities for farming arise in cleared forest areas.

There are two types of return migration that might have to be separately modelled. For the urban unemployed, the standard model expressions are valid; they too have to take into account the probability of finding an urban job. For those already employed in the cities, the decision is slightly different. Since they have a job, the probability calculation is not necessary; the rural wages would have to be absolutely greater than their current urban wages to induce them to migrate to the rural areas. In both cases we should, ideally, take into consideration the probability of finding a rural job, or some form of rural economic activity. This can probably be ignored; it is unlikely that anybody would return to a rural area unless they were almost certain that there was some way of supporting themselvés there.

With these modifications, the model would work in a satisfactory manner. For example, forest clearance due to logging can make subsistence farming in those areas much more attractive by reducing the required input effort and by opening up communications. Consequently, the net 'wage' a subsistence farmer receives rises; thus, provided this wage rise is sufficiently large, the model will predict a rise in migration to these areas.

There is one important problem with the mathematical model presented by Todaro in his Appendix 9.1 (Todaro, 1969). He shows that there is a critical (and very low) value of the elasticity of labour supply to urban-rural wage difference: above this critical value, job creation increases unemployment. However, there is nothing said about how this critical value can be calculated from empirical evidence; indeed, nothing is said about how migration changes as a function of expected wage difference. Another problem is that no guidance is given about how to model the rate of job creation; it is suggested that it might depend on urban wage rates and on other (unspecified) policy variables. On the other hand, one can parameterize the model in terms of the elasticity of labour supply to expected wage difference, and in terms of the number of jobs created per year per existing job.

Provided that one can find suitable parameters for the model, how can it be incorporated into the SARUM/TROPFORM/Land use Model programmes? The rather simple proposal for modelling subsistence agriculture (Parker, 1991) assumes exogenously given figures for the fraction of population in the subsistence sector, the yield per hectare obtained by subsistence farming, and the food production per capita in the subsistence sector.

The proposed migration model will be able to give figures for the proportion of the population in the subsistence sector. In order to produce these figures it is necessary to provide information on urban and rural wage rates, and on urban employment and unemployment.

Urban wages will come from SARUM; it simply has to be decided which sectors are predominantly urban (e.g. manufacturing, machinery) and to take an average value of the wage rate. There is an added complication in that the subsistence-sector modelling has to be done at the level of TROPFORM regions/countries and we only have wage rates for the more aggregated SARUM regions. We either have to assume that the wage rate is the same everywhere, or else that there is some fixed relation of the type 'wages in Cameroon are 40 per cent above average African wages'. Unfortunately, one does not have available good indicators of rural wages. The wages in the SARUM food sector correspond most closely to the wages paid in commercial agriculture, however, and will therefore be used for these purposes.

'Wages' in the subsistence sector can only be a notional figure inferred from the value of food produced and the labour input needed. In SARUM, we do have an exogenously provided figure for the per capita production of food in this sector. This production figure could be multiplied by the price of food generated by SARUM to give a measure of gross rural income. It is then necessary to subtract some measure of the 'costs' of production. If life for the subsistence farmer gets harder because of having to work on more marginal land, then the option of migration to the city becomes more attractive. Conversely, as mentioned above, opening up of the forests could have the reverse effect by decreasing these costs and thus increasing net rural income.

Conclusions

Although much research into migration has been done, no universally accepted theory has been developed. Indeed, there is some disagreement about what is actually happening in less developed countries. Is there excessive and growing urban unemployment? Is there a large pool of rural labour? To what extent can employment be characterized as formal or informal? What is the job search procedure used by migrants? In what way do incomes affect the decision to migrate? However, there does seem agreement that migration decisions are not irrational and are certainly affected by economic considerations. Furthermore, it seems generally accepted that demographic and educational characteristics do have an influence on who chooses to migrate.

There appear to be two ways of viewing migration which, very broadly speaking, can be represented by the approaches of Kannappan and Todaro. The central difference between them lies in the workings and efficiency of the labour market. Todaro, who follows in the tradition of Lewis's work (Lewis, 1954), assumes a relatively inefficient labour market in which neither wages nor supply and demand equilibriate. The alternative approach, inherent in Kannappan's article, is that there is a wide range of wages in urban areas and no clear distinction between formal and informal sectors. Furthermore, there is an extensive network of information flows which enables a relatively efficient means of directing labour to where it is needed. As a result, the wage differentials between cities and countryside, and the high urban unemployment rates are not as great as often assumed.

In terms of developing the model, it does appear that a separate labour model will be necessary for every TROPFORM region. The two approaches mentioned in the preceding paragraph lead to different formulations of the model: in one (Kannappan), migration is driven by job opportunities; in the other (Todaro), migration depends on expected wage differences.

Both models are capable of being implemented within the SARUM/TROPFORM framework and of reproducing the phenomena we wish to investigate. Because there are more unknown parameters and relationships in the Todaro model, it was decided to choose for the simpler job opportunities approach for the modelling of employment.

When incorporated into SARUM/TROPFORM, the migration model would endogenize the variable representing the proportion of the population in the subsistence sector. There are variables already available in SARUM that could act as inputs to the migration model; however, it will be necessary to make assumptions about whether the value of the SARUM variable is appropriate for all TROPFORM regions and countries within the larger SARUM region.

An alternative approach

Renaud (1979) shows that there is a strong relationship between per capita income and urbanization, based on a sample of 111 countries. Furthermore, urbanization is first accomplished by rural-urban migration, followed by internal growth of the urban population.

Although there exists a strong relationship between the levels of development and urbanization, the link between urbanization and industrialization appears to be much weaker. However, urbanization seems to be closely linked with industrial composition (change from traditional manufacturing to heavy manufacturing) rather than with industrialization.

Three methods are available for analysing migration and population distribution (IUSSP, 1980):

1) mechanical methods;
2) population change models (which are based on multi-regional cohort–component analysis);
3) analytical models (i.e., housing stock models).

The last two methods are theoretically superior to the mechanical models, but lack of data often rules out any application of complex models.

Several attempts have been made to specify complex models on migration. Wide-scale application, however, is not suitable. These models generally require very specific data input, which most countries that we are interested in are not able to provide. As a consequence, mechanical models have to be used, that do not describe the urbanization process extensively, but nevertheless can provide some rather realistic data about future urbanization rates.

To model the migration process in IDIOM, it is necessary to obtain detailed information on labour markets in every individual producing country. For an individual country, this is already very difficult to accomplish, due to, among others, limited data availability. Therefore it is not possible to give a detailed specification of the labour markets in each of the 34 individual countries.

As we have seen earlier, urbanization is strongly related to economic development. Regression analysis shows that a strong correlation exists between per capita income and urbanization. In our data set, based on UNDP's Human Development Reports, we have figures on both income per capita and urbanization. In Tables 6.1 and 6.3 some of these figures have already been presented for the tropical countries.

UN projections show that the absolute increase of the urban population from 1985 to 2000 is mainly due to the increase of the urban population in the developing regions (ca. 90 per cent). The urban population will increase with some 920m up to the year 2000.

United Nations projections

The United Nations have made estimates and projections of urban and rural populations (UN, 1980). These projections are based on the so-called 'United Nations method.' Basically this method involves extrapolating into the future the most recently observed urban/rural growth difference. This method results in a logistic time path of the proportion urban which has a peak velocity (annual absolute gain in proportion urban) at a proportion of 0.5 and has a maximum urban proportion, eventually reached by all countries, of 1.0.

The United Nations method

The urban/rural growth difference (URGD) is based on two censuses in each country. In these censuses the proportion living in urban areas, PU, has been measured. From these figures one can compute the URGD in the following way (UN, 1980, p.9):

$$URGD = \frac{1}{n} \, Ln \left(\frac{PU(2)/1-PU(2)}{PU(1)/1-PU(1)} \right) \tag{7.1}$$

Projections of the proportion living in urban areas in year t after the first census period can be calculated as follows (UN, 1980, p.9):

$$\frac{PU(t)}{1-PU(t)} = \frac{PU(1)}{1-Pu(1)} \cdot e^{URGD \cdot t} \tag{7.2}$$

The two census periods used in this study are 1975 and 1985. Although figures for 1980 are also available, we prefer to use a 10 year intercensal period. A five year intercensal period often results in inaccurate projections, due to the fact that there are rather big fluctuations in the urban proportions measured between two censuses. Based on these figures the UN has made projections up to the year 2025. These projections, however, are based on a modification of (eq. 7.2). Empirical evidence shows that the URGD tends to decline as the initial urban proportion increases. Our regression based on figures for 110 countries showed that the coefficient of correlation between the URGD and the initial proportion urban is -0.280.

The hypothetical URGD is (UN, 1980, p.10):

$$URGD_H = 0.044177 - 0.028274 \cdot \text{Initial proportion urban} \tag{7.3}$$

We therefore modified the URGD based on the two last censuses in such a manner that the last observed URGD was allowed to approach the hypothetical value increasingly during the projection period. In particular a set of linear weights were employed (UN, 1980, p.11) as shown in Table 7.3.

Table 7.3 Modified URGD weights

Projection period	Weight given to most recently observed URGD	Weight given to $URGD_H$
1990–1995	0.8	0.2
1995–2000	0.6	0.4
2000–2005	0.4	0.6
2005–2010	0.2	0.8
2010–2015	0.0	1.0
2015–2020	0.0	1.0
2020–2025	0.0	1.0

The above method does not seem to have a very strong theoretical background. First, an arithmetic approach has been developed which after close examination did not sufficiently correspond to the data. Because of this a new approach was developed in order to modify the first method. In the new approach a theoretical level of urbanization is compared to the observed level of urbanization, whereby the level of urbanisation is supposed to converge stepwise to the theoretical level during a period of 25 years. Moreover, a somewhat arbitrary stepfunction is introduced which does not seems to be supported by empirical evidence.

Therefore we also designed some alternative methods which were then applied in order to be able to compare the results with the UN projections, the latter being based on observations of two years: 1975 and 1985. The final projections are presented in Table 7.4, and graphically illustrated by Figures 7.9a-f.

We have applied the UN method described above on UNDP data on urbanization rates for two intercensal periods: 1960–1985 and 1975–1985. This has been done in order to compare the results with the United Nations projections, which are based on the intercensal period 1975–1985.

In Table 7.5 both the United Nations projection and the two projections based on UNDP data are presented. Table 7.5 clearly shows that the intercensal period applied has a significant effect on the projections of urbanization rates. Especially in Africa the differences between the three projections are rather large. This is due to the fact that the increase in the urbanization rate in Africa in the past decade is larger than in Asia and Latin America. Therefore it makes quite a difference which intercensal period is applied.

The question which one of the three projections is best cannot easily be answered. The United Nations projections are frequently used by

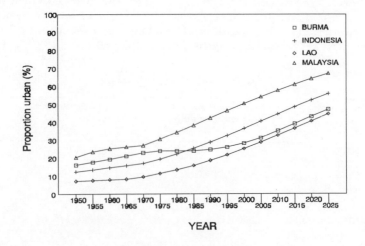

Figure 7.9a

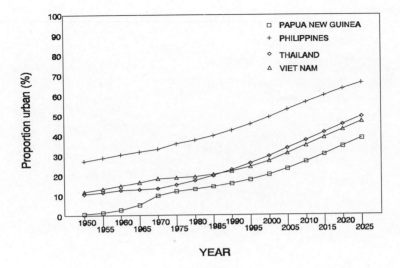

Figure 7.9b

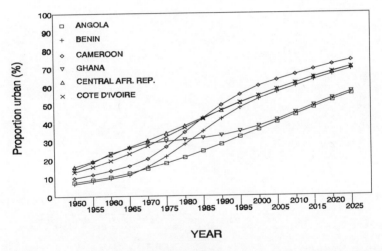

Figure 7.9c

modellers for scenario analysis. Therefore the choice for the United Nations projections seems to be suitable for our model. By using the United Nations projections, we will be able to compare our scenario results with other models. The differences between the model results will not be affected by using different projections of urbanization rates.

FINAL INTEGRATION

In Chapter 4 and the preceding sections of this chapter the three different models have been described which will be integrated into one

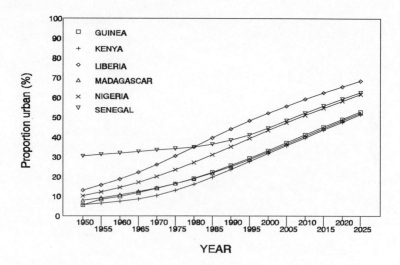

Figure 7.9d

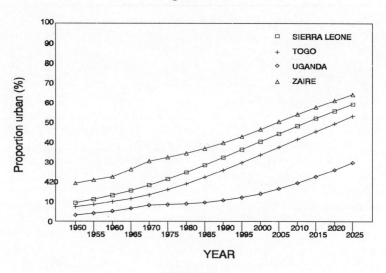

Figure 7.9e

model: IDIOM. These three models are: TROPFORM, SARUM and the Land Use model.

TROPFORM was the starting point of this project. It originally consisted of 21 producing countries and eight consuming regions (see also Table 4.1, p.72). We have extended the model to 34 producing countries and 10 consuming regions, which are listed in Figure 7.10.

The 34 producing countries in the new model cover the major part of the tropical rainforest (95 per cent). The choice of ten different consuming regions has been made on the basis of an existing dataset

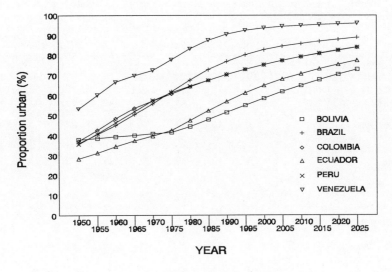

Figure 7.9f

Figures 7.9a–f IDIOM projections of urbanization rates, 1950–2025 (%)

for SARUM and the fact that within SARUM the countries belonging to a specific region should be as homogeneous as possible.

In Chapter 4 the shortcomings of TROPFORM were discussed. One of these shortcomings was that the demand for wood in the consuming regions was modelled rather simplistically. In the consumption module of TROPFORM (see pp.74–75) the demand for tropical wood depends on the population growth, the growth of income per capita, and the income elasticity for wood. The population growth rate, the growth rate of income per capita and the income elasticity of wood were set exogenously and are constant during the simulation period.

In order to refine the demand for tropical wood, SARUM has been integrated with the TROPFORM model. The demand for tropical wood still depends on the growth of the population, the growth of income per capita and on the income elasticity for wood. These variables instead of being set exogenously, have now been endogenized within SARUM, therefore we are better able to make predictions for future world exports of tropical timber. In SARUM, four different sectors have been specified:

1) Food
2) Natural products
3) Industry
4) Services

These sectors are the basic production units. For every simulation period, SARUM will provide production, consumption and trade flows for these four sectors. In the land use module, these figures are used for

Table 7.4 Urbanization rates in the tropical regions, projections to 2025

Country	Urbanization Rate (%)								
	1950	*1960*	*1970*	*1980*	*1990*	*2000*	*2010*	*2020*	*2025*
Tropical Asia									
Myanmar (Burma)	16	19	23	24	25	28	35	43	47
Indonesia	12	15	17	22	29	36	44	52	56
Lao Dem. Rep.	7	8	10	13	19	25	33	41	45
Malaysia	20	25	27	34	42	50	58	64	67
Papua New Guinea	1	3	10	13	16	20	27	34	38
Philippines	27	30	33	37	42	49	56	63	66
Thailand	10	13	13	17	23	29	37	45	49
Vietnam	12	15	18	19	22	27	35	43	47
Tropical Africa									
Angola	8	10	15	21	28	36	44	52	56
Benin	7	10	16	28	42	53	60	66	69
Cameroon	10	14	20	35	49	60	66	72	74
Centr. African Rep.	16	23	30	38	47	55	61	67	70
Côte d'Ivoire	13	19	27	37	47	55	61	67	70
Ghana	14	23	29	31	33	38	45	53	57
Guinea	6	10	14	19	26	33	41	49	53
Kenya	6	7	10	16	24	32	40	48	51
Liberia	13	19	26	35	44	52	59	66	68
Madagascar	8	11	14	19	25	32	40	48	52
Nigeria	10	14	20	27	35	43	51	58	62
Senegal	30	32	33	35	38	44	52	59	63
Sierra Leone	9	13	18	25	32	40	48	56	59
Togo	7	10	13	19	26	33	41	49	53
Uganda	3	5	8	9	10	14	19	26	30
Zaire	19	22	30	34	39	46	54	61	64
Tropical Latin America									
Bolivia	38	39	41	44	51	58	65	70	73
Brazil	36	45	56	68	77	83	86	88	89
Colombia	37	48	57	64	70	75	79	82	84
Ecuador	28	34	40	47	57	65	71	75	77
Peru	36	46	57	64	70	75	79	83	84
Venezuela	53	67	72	83	90	94	95	96	96

the food sector as inputs in order to derive commercial food production in the producing countries.

Furthermore figures from SARUM could be used to predict migration from one sector to another. For example, if the wages within the food sector become relatively unattractive compared to those in the other sectors (this is modelled within SARUM) it is very likely that people start to migrate from the food sector to the other sectors. SARUM provides relative wage rates, prices of commodities that will be used in the land use module.

As was mentioned earlier, a limitation of TROPFORM is that in the deforestation module (see also pp.77–78) no distinction is made between commercial and subsistence agriculture, and more generally, that deforestation was also modelled in a rather crude way. Therefore we have substituted the land use module for the deforestation module (which was already described on pp.183ff).

Table 7.5 Comparison of urbanization projections

Country	UN Projections			URGD 1960–1985			URGD 1975–1985		
	1990	*2000*	*2025*	*1990*	*2000*	*2025*	*1990*	*2000*	*2025*
Tropical Asia									
Indonesia	28.8	36.5	55.9	27.5	33.6	55.2	28.4	36.2	58.7
Lao Dem. Rep.	18.6	25.1	44.6	16.9	21.8	42.2	17.4	23.4	44.9
Malaysia	42.3	50.4	67.1	40.7	46.9	66.3	42.3	51.0	70.7
Pap.Nw Guinea	15.8	20.2	38.3	18.5	28.9	54.6	14.5	16.9	33.5
Philippines	42.4	49.0	66.1	40.9	46.1	65.0	40.5	45.2	63.9
Thailand	22.6	29.4	49.2	19.2	22.9	42.3	18.5	21.1	39.2
Vietnam	21.9	27.1	46.7	21.7	26.4	47.1	21.6	26.3	46.8
Tropical Africa									
Angola	28.3	36.2	55.6	29.3	38.7	61.9	29.1	38.1	61.1
Benin	42.0	52.8	68.9	42.5	56.0	76.5	45.8	63.7	82.9
Cameroon	49.4	59.9	73.9	49.8	63.0	80.7	52.3	68.3	84.9
Centr.Afr.Rep.	46.6	54.5	70.1	51.2	62.1	79.1	49.7	58.5	75.9
Côte d'Ivoire	46.6	54.6	70.2	54.4	69.0	84.8	59.7	78.9	91.4
Ghana	33.0	37.9	56.6	34.0	39.4	59.7	32.0	34.3	53.1
Guinea	25.6	33.2	52.9	25.4	33.1	56.0	23.1	26.7	46.5
Kenya	23.6	31.8	51.5	24.1	33.3	57.3	26.2	39.5	65.3
Liberia	44.0	52.1	68.4	45.6	60.4	79.9	41.9	51.6	71.6
Madagascar	25.0	32.4	52.1	23.3	29.1	50.7	22.6	27.3	47.9
Nigeria	35.2	43.3	61.6	32.9	39.7	60.9	30.5	33.6	52.9
Senegal	38.4	44.5	62.6	39.2	46.4	66.5	40.4	49.2	69.6
Sierra Leone	32.2	40.2	59.1	28.5	36.4	59.0	31.4	44.4	68.6
Togo	25.7	33.4	53.1	26.7	35.0	58.0	28.8	40.9	65.5
Uganda	10.4	13.8	29.6	7.5	9.2	21.1	6.5	6.7	14.8
Zaire	39.5	46.4	64.1	43.5	52.6	72.0	46.3	59.2	78.2
Tropical Latin America									
Bolivia	51.4	58.5	72.9	47.2	54.2	72.0	47.6	55.1	72.9
Brazil	76.9	82.7	89.0	77.4	83.7	90.9	78.4	85.5	92.2
Colombia	70.3	75.2	83.9	70.6	76.5	86.2	69.4	74.0	84.1
Ecuador	56.9	64.9	77.4	55.7	62.8	77.9	57.0	65.7	80.5
Peru	70.2	75.2	84.0	71.7	77.7	87.0	72.9	80.1	89.0
Venezuela	90.5	93.7	96.0	87.3	90.6	94.4	86.3	88.7	92.9

The land use module provides future trends of land use, giving us insight into how much forest is cleared for commercial agriculture and subsistence agriculture. This process has been described on pp.199–207.

Various methods of simulating rural-urban migration were discussed on pp.183–199. However, (lack of) available data has forced us to use migration figures from the United Nations, which are based on regression analysis of past trends of urbanization. These UN figures have subsequently been used to predict population figures of both the urban and the rural population. The result of the integration is a computer model, called IDIOM, that runs on any IBM compatible computer. The program is written in Turbo Pascal, and can easily be modified. The integrated model IDIOM has been developed in order to be used for scenario analysis. The model provides several options to see the implications of different policies for the tropical rainforest. In the

Producing countries

Asia	*Africa*	*Latin America*
Myanmar (Burma)	Angola	Bolivia
Indonesia	Benin	Brazil
Laos	Cameroon	Colombia
Malaysia	Central African Republic	Ecuador
Papua Nw Guinea	Congo	French Guyana
Philippines	Côte d'Ivoire	Guyana
Thailand	Gabon	Peru
Vietnam	Ghana	Surinam
	Guinea	Venezuela
	Kenya	
	Madagascar	
	Nigeria	
	Senegal	
	Sierra Leone	
	Senegal	
	Tanzania	
	Uganda	
	Zaire	

Consuming regions
North America
Europe
Developed Pacific
Tropical Latin America
Rest of Latin America
(formerly) Europe centrally planned
Tropical Africa
Rest of Africa
Tropical Asia
Rest of Asia

Figure 7.10 Timber producing countries and consuming regions in IDIOM

following chapter the results of several of these policy options will be discussed.

In Figure 7.11 the main interlinkages between the three modules – SARUM, TROPFORM and LAND USE – are summarized. The land use module is the part of the figure enclosed by the dotted line. The most important variables that are being transferred between the three modules are also specified in this figure.

DATA COMPILATION AND DATA SOURCES

For initialization of the model a comprehensive dataset had to be composed. To arrive at regional aggregates for the SARUM and TROPFORM regions as described in Table 7.1 and Figure 7.10, most data were collected on an individual country level. The base year 1990 was chosen, because this is the most recent year for which sufficient data were available. In the following sections the main variables for which data had to be collected and the compilation and sources of the dataset for these variables, will be described for each partial model.

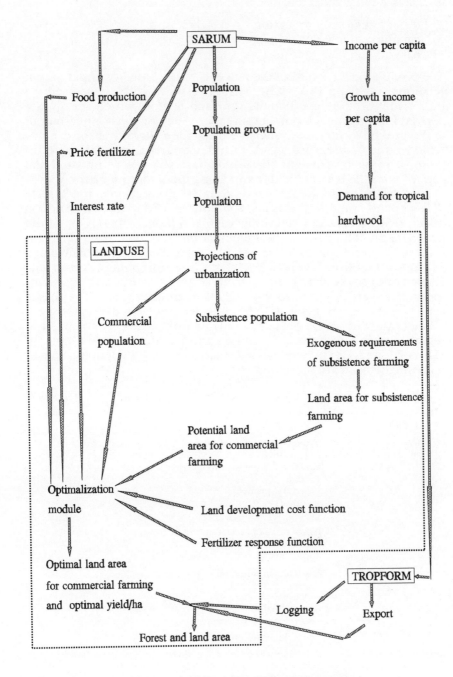

Figure 7.11 IDIOM: the basic model structure

SARUM

For initialization of the SARUM part of the model three main categories of data had to be collected:

1) data on regional production;
2) data on regional domestic demand; and
3) data on regional interlinkages.

The production structure of the model is partly based on input-output relations (see pp.180-182). These relations are expressed in intermediate input coefficients for each sector, which ideally are derived from regional input–output tables. Such regional input–output tables are, however, not available except for a limited set of country tables, none of them for 1990. In order to arrive at a fitting set of regional input coefficients, these country tables were updated and revised with the RAS method for updating input–output matrices.

For a discussion of the RAS method and its application to the SARUM data set (and other problems and solutions related to the construction of the SARUM dataset), see Hoogeveen and Blom (1989). One of the problems which arises is the incompatibility of data from different sources. To avoid this problem, use was made as much as possible of a single source (World Bank, 1992). Data from this source served as a framework for the data coming from all other sources; if necessary these data were adjusted to World Bank aggregates. Table 7.6 gives a survey of the data which had to be collected per main data category.

Since no ready-made data are available on the level of the SARUM regions, most data were collected on an individual country basis. In Tables 7.7 and 7.8 the regional aggregates of the production data and domestic demand are presented.

Tabel 7.6 Survey of SARUM Data Requirements

Data group	Required data
Production	Gross Domestic Product GDP growth Value added per sector: – Food – Natural products – Industry – Services
	For per capita production (and consumption) purposes Population Population growth
Domestic demand	Final consumption of: – Food – Natural products – Industrial products – Services
	Gross domestic investment
Regional interlinkages	Total exports per sector and per region of destination Total external long-term debt Official Development Aid per recipient and per donor region

Table 7.7 SARUM Production Data*

	Pop. (millions) 1990	Pop. growth % pa 1989–2000	GDP $m 1990	GDP growth % pa 1980–90	Value added 1990 ($m) food	nat. prod.	industry	services
N. America	277	0.8	5,962,350	3.4	52,631	79,420	1,821,464	4,008,834
Dev. Pacific	382	0.3	6,918,816	2.4	127,196	116,435	2,112,488	4,562,696
Europe	144	0.4	3,281,950	4.0	52,871	51,116	1,339,412	1,838,551
Trop. Lat. Am.	243	1.8	556,987	2.3	22,934	33,571	218,943	281,538
Other Lat. Am.	190	1.8	458,173	0.8	30,205	14,971	147,154	265,844
Eur. c. pl.	399	0.5	2,634,610	0.1	229,681	359,551	1,412,572	632,806
Trop. Africa	314	3.3	113,282	2.0	13,505	22,273	31,003	46,501
Other Africa	296	2.3	183,644	3.5	16,424	17,636	71,972	77,613
Trop. Asia	440	1.8	300,376	5.0	19,774	39,930	114,248	126,424
Other Asia	2439	1.8	1,289,872	4.9	109,738	155,880	477,802	546,452

* Population growth, GDP growth and sectoral distribution of GDP of small economies (population less than 1 million): average percentage regional income group.

Sources: Population: World Bank (1992, Table 1, pp.218–219, Box A1, p.285, Box A2 p.286).
Population growth: World Bank (1992, Table 26, pp.268–269, Box A2 p.286).
GDP: World Bank (1992, Table 3 pp.222–223); USSR, Libya, Vietnam, Myanmar, Yemen, Iraq and Liberia: WRI (1990, Table 15.1 pp.244–245).
GDP growth: World Bank (1992, Table 2, pp.220–221); USSR: estimate based on UN (1991, p.1974) and regional average, Libya, Lao, Myanmar, Jordan, Yemen: WRI (1990, Table 15.1, pp.244–245).
Value added per sector: World Bank (1992, Table 3, pp.222–223): USA, Canada, Ireland, Norway, Portugal, Spain, UK, Liberia, Yemen: WRI (1990, Table 15.1, pp.244–245); USSR, Libya, Myanmar, Malaysia: UN (1991, p.1975, p.1090, p.1203, p.1128 resp.).

Table 7.8 SARUM Data on Domestic Demand ($m 1990)*

	Food	Nat. prod.	Industry	Services	Gross dom. investments
N. America	499,426	62,894	930,904	3,512,634	982,484
Europe	799,719	81,067	1,052,303	3,566,908	1,515,724
Dev. Pacific	421,632	36,219	558,636	1,109,556	1,042,784
Trop. Lat. Am.	150,483	46,221	67,052	194,424	114,551
Other Lat. Am.	113,969	1240	100,512	186,220	78,664
Eur. c. pl.	692,860	33,558	351,592	840,607	711,884
Trop. Africa	43,108	1635	16,960	29,446	17,948
Other Africa	43,074	2726	35,558	75,122	42,106
Trop. Asia	79,670	4621	43,926	51,381	100,833
Other Asia	270,543	13,224	131,973	491,274	389,832

* Estimates for small economies and countries (population less than 1 million) for which no data were available, are based on averages of regional income groups.

Sources: World Bank (1992, Table 3, pp.222-223, Table 9, pp.234-235, Table 10, pp.236-237).

In the model, trade and aid are the key variables through which economic interaction between regions takes place. The 1990 interregional trade matrix per sector and per region of destination is presented in Table 7.9, with the exports on the rows and the imports in the columns of the matrix. Table 7.10, finally, gives the donated aid per region in 1990 as a percentage of regional GDP, the received aid per region as a percentage of total aid, and the long term Third World debt in 1990 per debtor and per creditor region.

TROPFORM

A full listing of the data requirements for the original TROPFORM model is given in Blom, Hoogeveen and Van der Linden (1990, pp.15-16). Since the deforestation module of the original TROPFORM model is replaced by the land use model to improve the mechanism to describe and simulate the linkages between agricultural development and deforestation, the data set was somewhat extended.

Some of the required data are provided for via the SARUM-TROPFORM linkages, such as population and population growth for all consuming and producing regions, GDP per capita and GDP per capita growth for the macro producing regions, and the agricultural production for each producing region (for the regional composition of the model see Figure 7.10). Although in case of the producing regions SARUM operates on a higher regional level of aggregation, the SARUM regional totals could be disaggregated to the producing country level, thanks to the fact that the SARUM regional aggregates are based on individual country data. For initialization of the model the most important additional data which had to be collected were the following:

- *Consumption module:*
  - total demand for tropical timber products in the base year for each consuming region.

- *Standing volume module:*
  - area of forest logged in the past forty year period.
- *Land use module:*
  - forest area per producing region;
  - farmland area per producing region;
  - annual increase in food consumption per capita per producing region;
  - yield per hectare for each producing region;
  - fertilizer consumption in each producing region;
  - urbanization rates for each producing region.

For the same reason as was specified with respect to the SARUM data, use was made as much as possible of a single data source. In the case of TROPFORM, this meant FAO data (production and trade yearbooks).

The data for the land use module have been presented in Table 7.12. For data on yield per hectare for the macro producing regions, the reader is referred to Table 5.4 and for individual producing countries to Table 5.5 and Hoogeveen (1992, Table 3 p.6, appendix 2 p.38, and appendix 4 pp.41-42). The data on urbanization were presented on pp.154-155, Tables 6.3 a and b, and on pp.202-203, Tables 7.4 and 7.5, of this study.

Table 7.9 SARUM International Trade Data ($m 1990)

Region of origin	Region of destination									
	North America	Europe	Dev. Pacific	Trop. Lat. Am.	Other Lat. Am.	Europe c. pl.	Trop. Africa	Other Africa	Trop. Asia	Other Asia
N. America		120,296	63,856	12,456	41,476	5640	1795	6668	11,649	61,004
food		8698	4617	901	2999	408	130	482	842	4411
nat. prod.		13,125	6967	1359	4525	615	196	728	1271	6656
industry		98,472	52,272	10,197	33,952	4617	1470	5458	9535	49,937
Europe	113,312		40,810	9503	16,951	51,075	16,120	36,557	14,302	80,477
food	8626		3107	723	1290	3888	1227	2783	1089	6126
nat. prod.	7896		2844	662	1181	3559	1123	2547	997	5608
industry	96,791		34,859	8118	14,479	43,628	13,769	31,227	12,217	68,743
Dev. Pacific	73,337	138,775		2404	6055	9616	1954	3978	17,329	59,452
food	2448	4632		80	202	321	65	133	578	1984
nat. prod.	2366	4478		78	195	310	63	128	559	1918
industry	68,523	129,666		2246	5658	8985	1826	3717	16,192	55,549
Trop. Latin Am.	23,155	16,889	4238		7693	1324	361	657	1000	3940
food	2412	1759	441		801	138	38	68	104	410
nat. prod.	3530	2575	646		1173	202	55	100	153	601
industry	17,213	12,555	3150		5719	984	268	489	744	2929
Other Latin. Am.	27,697	15,506	3728	4291		928	146	526	478	2615
food	7600	4255	1023	1177		255	40	144	131	718
nat. prod.	3767	2109	507	584		126	20	72	65	356
industry	16,330	9142	2198	2530		547	86	310	282	1542
Eur. c. pl.	3721	56,826	4231	445	424		628	2406	862	9550
food	223	3404	254	27	25		38	144	52	572
nat. prod.	349	5329	397	42	40		59	226	81	896
industry	3149	48,092	3581	377	359		532	2036	730	8083

Region of origin	Region of destination									
	North America	Europe	Dev. Pacific	Trop. Lat. Am.	Other Lat. Am.	Europe c. pl.	Trop. Africa	Other Africa	Trop. Asia	Other Asia
Trop. Africa	10,751	15,165	255	250	934	597		740	86	643
food	849	1197	20	20	74	47		58	7	51
nat. prod.	1400	1974	33	32	122	78		96	11	84
industry	8503	11,993	202	197	738	472		585	68	508
Other Africa	2833	30,743	2546	301	68	1303	639		614	2675
food	261	2830	234	28	6	120	59		57	246
nat. prod.	280	3039	252	30	7	129	63		61	264
industry	2292	24,875	2060	244	55	1054	517		497	2164
Trop. Asia	18,869	16,025	24,359	124	607	773	448	652		26,654
food	1750	1486	2259	11	56	72	42	60		2472
nat. prod.	3533	3001	4561	23	114	145	84	122		4991
industry	13,586	11,538	17,539	89	437	557	323	470		19,191
Other Asia	77,158	65,788	55,753	3601	4138	3730	3144	4480	22,244	
food	3525	3006	2547	165	189	170	144	205	1016	
nat. prod.	5008	4270	3618	234	269	242	204	291	1444	
industry	68,625	58,512	49,588	3203	3680	3318	2796	3985	19,784	

Sources: World Bank (1992, Table 14, pp.244–245, Table 16, pp.248–249), FAO (Table 5, pp.39–41, Table 6, pp.42–44, Table 8, pp.48–50), IMF (1992).

Table 7.10 SARUM: 1990 data on ODA and long-term external debt

	Donated aid as a % of GDP	Aid received as a % of total aid	Total external* long-term debt ($m)
N. America	0.233	0.00	-207,873
Europe	0.457	3.04	-349,246
Dev. Pacific	0.308	0.00	-377,972
Trop. Lat. Am.	0.003	2.82	72,871
Other Lat. Am.	0.000	8.27	180,959
Eur. c. pl.	0.000	0.10	89,609
Trop. Africa	0.011	17.06	106,390
Other Africa	0.006	32.25	126,232
Trop. Asia	0.000	10.73	122,608
Other Asia	0.489	25.74	236,603
Total aid ($m)	61,975		

* Sum of public and publicly guaranteed and private nonguaranteed long-term debt.

Sources: World Bank (1992, Table 19 pp.254–255, Table 20 pp.256–257, Table 22 pp.256–257), World Bank (1990).

Table 7.11 Standing volume module: area logged 1950–1990 (million m³)

	1950	1955	1960	1965	1970	1975	1980	1985	1990
Myanmar (Burma)	0.9	0.7	1.5	1.4	1.8	1.7	1.2	1.8	1.8
Indonesia	0.4	1.4	4.0	4.0	4.9	16.9	22.9	22.8	25.4
Malaysia	0.9	1.8	3.2	3.2	14.4	20.7	27.6	32.7	35.1
Papua New Guinea	0.0	0.1	0.1	0.1	0.4	0.9	1.0	2.0	2.5
Philippines	1.0	3.1	4.5	4.5	9.0	11.0	7.9	4.5	3.4
Cameroon	0.1	0.3	0.4	0.4	0.6	0.8	1.4	1.7	2.1
Centr. African rep.	0.0	0.0	0.1	0.1	0.2	0.5	0.3	0.3	0.2
Congo	0.3	0.2	0.3	0.5	0.7	0.8	0.4	0.5	0.7
Gabon	0.3	0.9	1.5	1.4	1.5	2.3	1.4	1.1	1.2
Zaire	0.3	0.5	0.8	0.4	0.1	0.6	0.3	0.4	0.4
Bolivia	0.1	0.1	0.1	0.1	0.1	0.1	0.2	0.2	0.1
Brazil	0.2	1.5	9.3	5.9	13.6	16.1	26.5	37.0	40.0
Colombia	0.1	2.2	2.2	2.2	2.3	2.6	2.1	2.0	2.0
Ecuador	0.1	0.2	0.6	0.8	1.1	1.7	1.8	2.0	2.3
French Guyana	0.0	0.0	0.0	0.0	0.1	0.0	0.1	0.2	0.2
Guyana	0.1	0.2	0.2	0.2	0.2	0.2	0.1	0.2	0.2
Peru	0.1	0.1	0.1	0.2	0.6	0.6	1.2	1.2	1.1
Surinam	0.2	0.1	0.1	0.2	0.2	0.2	0.2	0.2	0.2
Venezuela	0.2	0.0	0.3	0.3	0.4	0.4	0.6	0.6	0.6

Source: FAO yearbook of forest products, various volumes.

Table 7.12 TROPFORM: data deforestation module

Producing region	Food cons. p.c. av. ann. gr (%) 1979–89	Forest area in 1000 ha 1989	Farmland in 1000 ha 1989	Fertilizer cons. (100 gr. nutrient/ha) 1970–71	1989–90
Tropical Asia					
Myanmar (Burma)	0.87	32,418	10,395	21	86
Indonesia	1.56	113,433	33,060	133	1166
Laos	0.88	12,800	1701	2	3
Malaysia	0.12	19,100	4907	436	1572
Papua New Guinea	0.88	38,230	472	58	399
Philippines	−0.15	10,550	9210	287	674
Thailand	0.11	14,240	22,896	59	365
Vietnam	0.88	9800	6935	513	841
Tropical Africa					
Angola	−0.07	52,950	32,600	33	74
Benin	0.71	3520	2302	36	18
Cameroon	−0.11	24,650	15,308	34	41
Centr. African Rep.	−0.48	35,810	5,006	12	4
Congo	0.85	21,180	10,168	525	32
Côte d'Ivoire	−0.07	7630	16,660	74	113
Gabon	−0.07	20,000	5152		27
Ghana	1.80	8140	7720	11	31
Guinea	−0.28	14,640	6878	44	11
Kenya	−0.32	2360	40,528	238	481
Madagascar	−1.49	15,680	37,092	61	36
Nigeria	−0.01	12,200	71,335	2	121
Senegal	−0.20	5942	10,926	17	55
Sierra Leone	−1.53	2070	4005	17	3
Tanzania	−0.22	41,060	40,250	31	93
Uganda	0.15	5610	8505	14	1
Zaire	−0.29	174,640	22,850	6	10
Tropical Latin America					
Bolivia	−0.76	55,650	30,110	7	23
Brazil	0.09	551,330	248,650	186	430
Colombia	0.41	50,600	45,680	287	902
Ecuador	1.15	11,200	7753	133	338
Fr. Guyana	1.43	7300	18		
Guyana	1.47	16,369	1725		
Peru	0.23	68,650	30,850	300	411
Surinam	1.87	14,855	88		
Venezuela	−0.24	30,465	21,545	170	1507

*Estimate based on daily calorie supply per caput. Weighted regional average for countries for which no data were available: Laos, Papua New Guinea, Vietnam, Angola, and Gabon.

Sources: FAO (1990, Table 1, pp.3–14 and Table 106, pp.237–238), World Bank (1992, Table 4, pp.224–225).

PART III

SCENARIOS

8

SCENARIO ANALYSIS AND MODEL SIMULATIONS

INTRODUCTION

After having discussed IDIOM's characteristics, the issues: 'What are the main causes of deforestation in the humid tropics?' and 'What policies will reduce this deforestation?' can be analysed with the help of the model. There are, in addition, important subsidiary considerations such as whether the policies are feasible, how much they will cost, who will pay and who will benefit.

We wish to set our investigations within the framework defined by these questions. However, we face the problem that always arises when looking at future policy: there are innumerable possible 'futures'. Not only are there many uncertainties about human social and economic systems that will evolve, but we also have a choice about which paths to take; to use the terminology of decision analysis (Dennis and Dennis, 1991), there are many 'states of nature' over which we have no control, and many 'decision alternatives' from which to choose.

We shall organize these multiple futures by defining a small number of scenarios, each of which will integrate policy and assumptions about future developments to give a consistent picture of the world. Details associated with each scenario will be derived using the IDIOM combined global and forestry model.

THE MAIN COMPONENTS OF THE SCENARIOS

As mentioned above, we are interested in the causes of deforestation and possible policies to reduce such damage. Although mankind uses the resources of tropical forests in many varied ways, two particular kinds of exploitation are of the greatest importance (Amelung and Diehl, 1992): logging and clearance for agriculture. If we extend agriculture to include population growth and internal development, then we can use these two categories to define one dimension of the scenarios. They are differentiated by the fact that the logging category is concerned directly with the extraction of trees, whereas agriculture and development deal with the acquisition, for other purposes, of the land on which the trees stand. This dimension could also have overtones of 'international' versus 'domestic' since logging and timber sales have been traditionally associated with international trade, although domestic demand is assuming an ever more important role (as discussed in the LEEC (1993) report for the ITTO). In contrast, 'agriculture and development' can be considered primarily as a description of what is happening within a country.

When looking at the logging side of a scenario, ultimately one has to consider the level of demand for tropical wood and derived products. The domestic demand for these is very much connected with the continuing development process and thus is unlikely to be amenable to changes in consumption patterns. However, the demand from the advanced industrialized countries (AICs) has come under scrutiny (for example, the original Netherlands' policy of importing only sustainably managed tropical timber by 1995). Indeed, according to Friends of the Earth,[1] 58 per cent of those surveyed claimed they would not buy a timber product if they knew it came from a rainforest. Thus changing consumer preferences in advanced industrialized countries would have a direct influence on tropical timber consumption. Such a change will be considered as an independent scenario.

A dominant theme in a recent LEEC report (LEEC, 1993) is that policies have to be considered in the light of whether they encourage the development of sustainable forest-management practices. Thus, as a counter to a direct reduction in consumer demand, a scenario should be considered in which there is a move to sustainable management. Such practices, inevitably, cost money and therefore it is important to consider where the finance is to come from. Whether this source is development aid or a tax on the tropical timber trade gives another dimension to this scenario.

Additional scenarios can be developed by moving to the indirect impact of agriculture and development on deforestation. The issue of development is very wide and has been extensively analysed in the literature (see Chapter 6). Our concern here is with the use of forest lands for other purposes. Although, in some countries, mineral extraction and the building of dams can lead to the loss of considerable areas of forest, our main concern is with people clearing the forest in order to grow food. This, of course, can happen in many ways, through subsistence farming, the development of commercial arable farming, the creation of cattle ranges and the planting of 'cash crops' such as palm oil. In addition, other issues are raised such as property rights, traditional communal practices and rural–urban migration.

However, underlying all our analyses is the fact that people need to eat and that food production needs land. One essential aspect of our scenario must be population policy, essentially by means of birth control and resettlement, but remembering that other policies such as overt industrialization will directly affect the rural–urban division (Parker, 1992). Scenarios are performed that investigate the sensitivity of the model to different assumptions about rural–urban migration.

Agricultural development is at the heart of any development process. Increases in yields (e.g. the 'green revolution') will, for a given population, result in less land being needed for food production, thus

---

[1] Press release 28 March 1992.

lessening the pressure on the forests. The only realistic source of funding for this development-oriented scenario is aid from the AICs.

An important debate in the field of development centres on 'aid or trade?' It is often argued that policies that open markets to Third World exports help those exporting countries far more than any donation of aid. The final scenario will thus be based on 'export-led Third World industrialization'. At this stage, it is impossible to tell whether the shift to a higher-technology agriculture and lower rural population will lead to less demand for forest areas given that there will, inevitably, be a greater domestic demand for wood and wood products.

THE SCENARIOS

In the light of the above discussion, we can define the scenarios more precisely. The components of the scenarios are shown clearly in Table 8.1.

Table 8.1 The specification of the scenarios

International finance transfers and trade population	Demand for wood/Logging practices		
	No change	Change in consumer preferences	Sustainable logging
No change	Base scenario	Scenario E	
Tax transfer for sustainable logging			Scenario A
Aid finance for sustainable logging			Scenario B
Aid finance for sustainable development and population measures	Scenario C		
Trade changes: export-led Third World industrialization	Scenario D		
Increase in urbanization	Scenario F		
Decrease in urbanization	Scenario G		

Having specified the broad outlines of the scenarios allows us to look in more detail at their implications and at the computer inputs necessary for realizing the scenarios as computer-based simulations.

The base scenario

In scenario analysis, one gains insight not from the absolute consequences of any one scenario, but from the changes that occur between a set of scenarios. By this means it is possible to infer certain causal relations; the introduction of a particular policy causes certain identifiable consequences.

Furthermore, with the help of a computer simulation model, these consequences can be quantified. It is therefore necessary to have a starting point from which all changes are measured and so can be considered as a yardstick against which all policies are measured.

A dual model (global model and forestry model) such as used here requires a very large amount of data before any experiments can be performed. Much of this effort is described in the previous chapter. As much information as possible is gathered for the base year 1990 to ensure that the model results are correct at the start of a simulation. Once a simulation into the future is commenced there is, of course, uncertainty. The purpose of a scenario is to simplify the description of that uncertainty by reducing it to a few 'principal components' which can be expressly chosen and controlled.

In both SARUM and TROPFORM the driving variables that determine the broad future paths are the growth of population and the economic growth as characterized by rates of increase of the GDPs in the various regions. These two assumptions together give the growth in per capita income, a useful measure of standard of living and, more indirectly, of technological progress. It is sensible for a base scenario to represent a continuance of current trends and to be in accord with a consensus view of important agencies connected with future world development. We have, therefore, based our projections of population and GDP on those used by the World Bank (1992). It must be remembered that although these have an official imprimateur they do not represent a certain outcome, or even necessarily the most likely outcome. The role of these projections is to provide a point of comparison, and generally acceptable view of the future.

Scenario A. Sustainable forest management: tax transfer financed

Description
At the core of this scenario lies the assumption that commercial forestry in the tropical countries will be managed in a sustainable way. However, this does raise serious questions as to what is meant by 'sustainable'. It can be argued that any anthropogenic interference changes the ecosystem in some way; thus the original forest ('virgin' forest or climax ecosystem) is not sustained. In this context, such a view is probably too extreme as it implies that the only sustainable logging system is to carry out no logging whatsoever.

It is certain, however, that any logging regime that involves a fall in the area of commercial reserves is not sustainable. What is wanted is a management system which involves replanting and then harvesting on a continuous cycle so that the age profile of the trees over a suitably large area remains the same (WRI, 1990).

The other side of this scenario relates to the financing of sustainable management. It is clearly very difficult to make precise estimates of the costs of sustainable forest management. Not only is it difficult to estimate the cost of any action within a programme, but the actual content of the programme is hard to define. Ferguson and Muñoz-Reyes

Navarro (1992) mention annual costs for sustainable management lying between $300m and $1500m (in 1990). These figures are discussed further in the LEEC report (1993) where it is also concluded that a tax of 1–5 per cent would have little distortionary effect on the market, but that its revenue would probably fall short of the amount needed. These figures should also be viewed in the light of various estimates produced in Agenda 21. The programmes described in Agenda 21 cover broader areas than simply sustainable commercial logging; details can be found in Robinson et al. (1992). The programme that explicitly includes sustainable logging, as well as reforestation, would require an annual contribution on grant or concessionary terms from the developed world of $3700m (towards a required total of $10,000m). This total figure is corroborated by considering the rate of forest loss and replanting costs. The average annual rate of tropical deforestation during the 1980s is estimated as 12.0 m ha (WRI, 1990). If this figure is combined with estimates of replanting costs of $700–1000 per ha (World Bank, 1990a), annual replanting costs would amount to $8400–12,000 m. However, both these approaches include considerable expansion and rehabilitation of forest areas and thus would lead to a considerable overestimate of the costs of sustainable logging alone. Furthermore, it is admitted in Agenda 21 that these estimates should only be considered as very broad order-of-magnitude figures. Considering that the Agenda 21 estimates are high, but approximate, it was decided to use the highest figures mentioned in studies for sustainable logging alone, i.e. $1500m per year.

In this scenario, a tax will be raised as a percentage of the value of consumption of tropical hardwoods in the advanced industrialized regions and will be of sufficient size to generate a flow of $1500 million per year over the course of the simulation. Such a tax will have an effect on demand, but this can be estimated through demand elasticities. Indeed, it must be noted that the tax will have to be high enough to counter this reduced demand. One final aspect of this tax is that it constitutes a financial flow from the developed world to certain less developed countries and thus is analogous to aid. However, it would not be an explicit part of the aid budgets of the donating countries and would be channelled into investment in the forestry sectors of the recipients.

This move to sustainable forestry implies very much more active management, with a long-term view being taken. Thus the time horizon has to be commensurate with the regeneration time of the trees (40 years). Maximum logging rates will have to take this longer view into account. On the other hand, the active management of reserves will raise the yield of regenerated forests which is, at present, assumed to be only one half of the initial yield.

Finally, it should be noted that a transition to sustainable forestry cannot happen overnight. Indeed, it will be impossible to implement a fully sustainable system until at least one complete cycle of planting and harvesting has been completed; i.e. it will take at least 40 years.

Quantitative assumptions and implications

The main assumption simply is that replanting of commercial reserves will be such that the total area of commercial reserves remains constant. However, it must be remembered that replanting does not immediately produce mature trees ready for harvest. The model will, therefore, have to keep account of the fact that the land has been allocated to forestry and cannot be used for any other purpose. Then, after a suitable regeneration time (40 years), these areas will be ready for harvesting.

We shall assume that the new policy is implemented immediately (1990). The time horizon for calculating maximum permitted logging rates is essentially a matter of attitude, and thus it can be changed to 40 years from 1990. Raising the yield from regenerated forests depends on new management practices. Although newly logged areas can be managed in a sustainable manner right from the start, it will not be possible to effect major changes on forests that were logged 10 or 20 years ago. As a result, the raising of regenerated yields from 50–100 per cent of initial yield can only be brought in over a complete tree-growth cycle, i.e. 40 years.

Simulating the financing of this policy will have to be performed iteratively. The first stage will be to run the model with replanting included. The annual average value of exports to the AICs over the simulation period will be calculated. These provide initial estimates of the imports from the three advanced industrialized regions (M_{0i}, i=1..3, in \$m/year). Knowing the three demand elasticities it is possible to calculate the necessary tax level. If the fractional tax rate is T, and the desired average revenue from the tax is \$1500 million per year, then we can write down the equation:

$$\sum_{i=1}^{3} TM_{0i}(1 + T)^{e_i} = 1500$$

This equation in T is relatively easy to solve, either through numerical techniques, or by trial and error in a spreadsheet.

The new tax is then applied to the demand elasticities to calculate the reduction factor in demand for each importing region. The model can then be run again. However, the demand by the AICs is, in some circumstances, modified within the TROPFORM part of the model. When commercial reserves fall to a level such that they would be exhausted within the time horizon, then logging activity is restrained. This, in effect, means that there has to be some rationing of the tropical hardwood. The assumption within the model is that domestic consumption within the producing countries has priority over exports. Consequently, a tax on exports will not lead to as big a reduction in trade as expected; the fall in demand will lead to less rationing being necessary, hence some of the cuts in demand due to rationing will be restored. It is, therefore, necessary to iterate on the tax rate so as to raise the desired revenue. However, only one further simulation is used as an interpolation between the zero tax rate and the rate calculated from the demand equation above is sufficiently accurate.

We have assumed that tropical countries are able to finance sustainable forest management by a tax on consumption in the AICs. Although it is not an explicit aid transfer, such a tax does represent a flow of resources from developed to less developed countries. Its financial effects on balance of payments, and its support for forestry development, lead to this tax having very much the same impact on the recipient countries as an aid project (However, it must be remembered that the tax will affect demand for timber products in a way quite unlike an aid donation.) These international transfers should, therefore, be represented in SARUM.

The extra annual flow of $1500m is assumed to come from the three advanced industrialized regions (North America, Europe and Developed Pacific) and to be divided among them in proportion to their average imports of tropical wood. This is achieved by their donating an extra small percentage of their GDPs; the percentages are fixed so that the total average donation over the period of the simulation is the desired $1500m per year.

Scenario B. Sustainable forest management: aid financed

Description
This scenario is the same as Scenario A except that the costs of sustainable management are met by aid transfers. This has a very important consequence: the demand for imports of tropical hardwoods is unaffected by any price-elasticity effects.

Quantitative assumptions and implications
Again, these are the same as for Scenario A but rather simpler. Replanting, time-horizon changes and the yield changes of regenerated forests are implemented exactly as before. However, no changes have to be made to demands.

In contrast, there is a change that has to be made to the SARUM part of the model. The replanting costs are worked out from an initial simulation and, in a second run, are introduced as aid flows. Existing aid is simply raised by a suitable fixed proportion as to generate the desired additional flow. The receipts are then apportioned between the tropical regions in proportion to their production of timber.

Scenario C. Sustainable development: aid financed

Description
There are many aspects of development, of which environmental degradation is only one. However, it is now realized that there are many interlinked problems and that a narrow focus on particular issues is not the most fruitful approach, an idea which, indeed, is a continuing theme of this report. We can also note that environment is playing an ever more important part in the thinking of aid agencies. For example, the Development Assistance Committee of OECD established a

Working Party on Development Assistance and the Environment in 1989 (OECD, 1991) which has been followed by a joint meeting of Environment and Development Ministers, as well as producing detailed study papers for Ministers and heads of aid agencies.

On a larger scale, the World Bank devoted its 1992 Development Report to the subject of development and the environment (World Bank, 1992). Indeed, some of the suggested policies in this report form the basis of this scenario, but also contribute to the other scenarios described here. This report is very wide-ranging and covers many issues apart from deforestation (e.g. soil erosion, water pollution, industrial emissions and the greenhouse effect). It also reiterates the view expressed on pp.12–17 and reconfirmed in the 1993 LEEC report that logging per se is not the most serious cause of deforestation.

Since deforestation must be seen as a complex socio-economic phenomenon, the aim of this scenario is to investigate whether more general developmental policies are more effective than policies specifically targeted at forest management. In particular, we are interested in the effects of combined industrial, agricultural and population policies. There is clearly a prima facie case that higher yields per hectare and lower populations will reduce pressures on the forests. The World Bank Report has several chapters that look at specific environmental issues, but they are all drawn together in the final chapter. These are summarized in the next subsection and seven particular policies are assumed to be implemented in this scenario.

Quantitative assumptions and implications

As mentioned above, this assumption is based on various policy analyses of the World Bank; their programmes and cost are detailed in Table 8.2.

Although programmes 1 to 5 are of great importance in reducing environmental impact, in this scenario we are going to concentrate on the four programmes 6–9 which are particularly relevant for

Table 8.2 World Bank sustainable development programmes: annual investment costs

Programme	*Additional annual investment in 2000 ($ billion)*
1) Water and sanitation	10.0
2) Control of particulate emissions from coal burning	2.0
3) Reducing acid decompositions	5.0
4) Changing to unleaded fuels and controls on vehicles	10.0
5) Reducing emissions, effluents and waste from industry	10.0–15.0
6) Soil conservation, afforestation, including training	15.0–20.0
7) Additional resources for agricultural and forestry research	5.0
8) Family planning	7.0
9) Female education	2.5

agriculture and population. The total annual cost lies in the range $29.5–34.5 billion.

We shall assume that $32 billion per year will be invested in the less developed countries and that the source of the finance comes from aid (with the same percentage rise in all donor countries). In addition, it will be distributed to the recipient regions in the same proportions as at present. It would not be possible to introduce such far-reaching aid programmes in a very short time; the World Bank envisages the necessary investment levels being achieved by the year 2000. Thus we assume that the extra aid is phased in over the 1990s until the full $ 32 billion is achieved. Thereafter, we take the somewhat generous view that the aid will rise in line with the donors' GDP growth.

It is also necessary to consider into which sectors of the SARUM model this additional aid-financed investment will flow. The contributions to education and family planning belong in the Services sector whereas the remainder goes to the Agriculture and Natural Products sectors. It is then assumed that the investment is divided between these two sectors in proportion to their base-year outputs. These macroeconomic effects will be implemented in the SARUM part of the model as described.

However, the consequences of the aid programmes for the forestry (TROPFORM) part of the model do have to be clearly and quantitatively specified. One natural consequence is that there will be a commitment to sustainable management of forests as in Scenarios A and B. The investment in agriculture will be represented by an additional 1 per cent per year growth in agricultural yields. Such scenarios have been investigated by Hoogeveen (1992). In addition, this rate is not inconsistent with the analysis of over 200 studies (Doolette and Smyle, 1990). (Over a 35 year period a 1 per cent annual growth rate will result in about a 40 per cent rise in yield.) Finally, population is taken to be subject to a rapid decline in fertility as assumed by the lower projections of the World Bank[2]. This is simplified to the assumption that growth of population in the developing regions up till 2025 is two-thirds of that in the base scenario.

Scenario D. Third World industrialization: trade induced

Description
The scenarios considered so far have effected transfers from the richer to the poorer regions of the world either through a tax on trade, or directly by means of aid. The impetus behind this scenario is that a freeing of trade can be just as effective a way of accomplishing such an aim. There is a considerable literature on this topic; see for example Johnson (1967), Thirlwall (1976, 1983), Cohen (1986) and IOV (1990). It is also important to note the work of Yeats (1982) which contends that the stimulus to development from increased trade opportunities has, in some periods, exceeded that due to aid. The importance of trade has also been recognized in an earlier study carried out using the

---

[2] See Figure 1.1 of the World Development Report, 1992.

SARUM model (Jepma, 1992). Here it was calculated that the point elasticities of GDP per capita to increased trade lay in the range 0.1 to 0.3, with the greatest gains in welfare occurring within the poorest developing regions.

The importance of international trade is also acknowledged in the World Bank's 1992 Development Report. They state that a halving of tariff and non-tariff barriers in the main industrial countries would raise an additional $65 billion for the less developed countries by the end of the decade, a figure comparable with the cost of the investment programme for sustainable development described in Scenario C. Although this assumption is central to the quantification of this scenario, it must be remembered that the money received from additional trade follows a very different path from that given as aid.

An increase in trade in industrial products brings additional revenue to the sectors of the economy that produce these goods. Much of this will be used for expansion, both in terms of investment and labour force, and will have a stimulatory effect on the economy as a whole. However, there is no reason to believe that such additional revenues will be channelled into specific development projects, or into programmes to aid sustainable environmental management, or to social programmes such as birth control and female education. One must ask, therefore, whether an enhancing the export opportunities of the less developed regions will affect environmental quality. It is clear that a stimulus to exports of industrial products will alter an economy's structure. One would expect to see a greater expansion in industry than in agriculture or services. Almost inevitably, a larger industrial sector means a higher urban population; i.e. rural–urban migration will be encouraged as job opportunities are created in industry (Kannappan, 1985). Although urbanization brings its own environmental problems (such as sanitation and water pollution), it can have a beneficial effect on forest encroachment (see page 29 of the World Development Report 1992 published by the World Bank). There are fewer people living primarily by subsistence farming in the vicinity of the forest, consequently there is a greater proportion of the population whose food originates from commercial agriculture. Commercial agriculture has a higher yield; hence, on balance, less land is needed for food production.

Quantitative assumptions and implications
The World Bank refers to a halving of trade barriers by the end of the decade (op. cit., p.175). However, in a simulation model one has to specify precisely what quantity is being halved. SARUM has a system of 'trade biases' that represent all the non-price barriers to trade (distance, culture, political relations etc.) (Parker, 1977). When barriers against imports from the developing countries are referred to, what is meant is that there are obstacles to trade compared with trade coming from other advanced industrialized countries. In this scenario, therefore, the bias of the advanced industrialized regions against imports of industrial goods from the 'South' must be reduced; furthermore, they must fall

towards a bias that is typical of trade between the advanced industrialized regions of the 'North'. It is assumed that the gap between the initial biases and this typical 'intra-North' bias halves between 1990 and 2000. The idea behind this scenario is that there is a reasonably gradual transition towards a world where the South is on an equal footing in relation to access to the markets of the North. Consequently, it is assumed that this fall towards the lower target does continue at the same rate into the twenty-first century.

Apart from the changes in trade relationships, nothing else is changed compared with the base scenario. It does not seem reasonable to suppose that the extra export sales of the industry sector will be channelled either into agricultural investment or social programmes for birth control and education. Thus agricultural productivity and population growth will remain unaltered. However, as mentioned above, there will tend to be a movement of labour into the industrial sector. The percentage rise observed compared with the base run will be used to modify the urban population pro rata. The subsequent reduction in the rural population will reduce the land needed for subsistence agriculture. On the other hand, this reduction will be partly offset because the area dedicated to commercial agriculture will rise in response to the larger urban population.

Scenario E. Change in consumer preferences

Description
The philosophy behind this scenario is that of 'green consumerism' in the advanced industrialized countries. Following the analysis in Annex D of the LEEC report (Bishop and Bann, 1992), there is clearly some willingness on the part of the British public either to pay more for products made from tropical hardwoods, or even to refuse to buy such products at all. The most extreme reaction reported is that 58 per cent of people would not buy a product they knew to come from a rainforest.

Also reported are the proportions of customers willing to pay prices that are higher than current values by a specified amount in order to obtain timber products that are, in some, way, 'environmentally friendly.' It appears that somewhere between a quarter and a half of customers would tolerate a price rise of 10 per cent. The implication is that between a half and three-quarters would not be prepared to pay even a small increase in order to support producers of wood from sustainably managed forests. However, one can infer that if wood prices were to rise because of more expensive forest-management practices, then demand would fall quite severely, a deduction that is corroborated by the very high elasticities of substitution for source of origin (Constantino, 1988). This elasticity estimate does rely on there being other suppliers whose prices have not changed, an assumption probably implicitly made by the respondents to the surveys on tolerable price rises (see also pp.40–54).

Although it appears that there are some contradictions in people's professed attitudes to questions on purchasing intentions, it is reasonable to assume that there are environmentalist sentiments among purchasers. We have already mentioned the Friends of the Earth Survey and the Netherlands' policy target of ceasing to import wood from non-sustainably managed forests. Parallel examples have already been seen in the greatly increased market share of non-polluting cleaning products in countries such as Germany (where they have more than half the market).

Quantitative assumptions and implications

It is difficult to give a realistic quantitative estimate of the potential fall in demand due to changed consumer preferences. A scenario not only has to be a realistic and self-consistent picture, but also has to highlight the consequences of different futures by providing distinct contrasts. Differences in scenarios do, therefore, have to be quite strong. Consequently, we will assume in this scenario that per capita demand for tropical hardwood in the AICs falls to half its 1990 value.

There is another dimension to this scenario; that is the speed of this fall. An order of magnitude can be inferred from the adjustment lags reported in the elasticity studies of Buongiorno and Manurung (1992) which vary between 4.1 and 6.7 years. These too relate to changes in consumer preference, but those caused by a change in price rather than in attitude to the product. However, these figures do indicate how quickly consumers can adjust once they have the willingness to adjust. One other point that needs to be noted is that these figures do not represent the time to make the complete adjustment. For example, a time lag of five years indicates that the gap between the current position and long-term intentions falls by 1/5th, or 20 per cent, every year, thus implying that this gap over a longer period of time closes according to a negative exponential function. Taking this into consideration, a reasonable assumption is that the 50 per cent fall in demand will take two lag periods, or 10 years, to occur. That is, the component of per capita demand that is not related to income will halve between 1990 and 2000 and stay constant thereafter. As income rises, however, there will still be a certain propensity to increase the consumption of tropical hardwoods because of the positive income elasticities. This rise will, in part, be due to people lower in the income distribution becoming able to purchase certain goods that they could not previously afford; that is, the increase is due to a relaxing of budget constraints rather than an intrinsic change in attitude to the purchase of these goods.

Scenarios F and G. Sensitivity to changes in urbanization

Description

As mentioned earlier in relation to Scenario D, it is thought that urbanization might have an appreciable effect on land use. Discussion of various migration models in Chapter 7 highlighted the problems of finding an agreed robust model and making accurate predictions. These scenarios, unlike the others, are not focused on a particular policy options, but serve to help analyse a very important aspect of development that is critical to deforestation. However, if urbanization is a critical issue, then any policies that have significant effects on the rural–urban division of population would have to be analysed in terms of their environmental consequences.

Two scenarios are analysed: Scenario F increases urbanization, thus implying more people in the rural subsistence sector, and Scenario G posits a decrease in urbanization, the size of the change being comparable to that in Scenario F, but in the opposite direction.

Quantitative assumptions and implications
The methodology of the urban-population projections is described on pp.196–199. In essence, the urban populations rise continually, indeed, in the very long term, moving towards a 100 per cent urbanization. These scenarios change the gap between the projected percentage of population in the cities and 100 per cent. In Scenario F it is assumed that this gap decreases by a quarter (25 per cent) whereas in Scenario G it increases by a quarter. There is no special significance in these figures; they are simply an implementation of the scenario-analysis technique where it is hoped that changes are not beyond the realms of possibility, yet are of sufficient size for the impact to be appreciable and conclusions drawn. Since the urban populations in the base year are known, it would not be reasonable to make large changes immediately. It has therefore been assumed that these changes are brought in gradually over a 35 year period.

As yet, the migration modelling within IDIOM has not yet been developed to a high degree of sophistication (pp.183–196 demonstrate very much the challenge to researchers). For example, the differences between the rural non-subsistence and the rural subsistence sectors of the population has not been analysed in detail. At the moment, therefore, the inferences drawn about the amount of forest land lost to subsistence and commercial agriculture assumes the apportionment of new land used by subsistence farmers between forest and non-forest areas remains constant, the calibration being based on estimates such as those produced by Myers (1989) and Otto (1990).

SIMULATION RESULTS AND COMPARISON OF THE SCENARIOS

This section will present the results of the computer model and discussion of the policy implications that arise from the scenario analysis. Pages 230–250 analyse the Base Scenario in comparison with the various policy scenarios. This analysis concentrates on the larger scale issues, mainly from a socio-economic viewpoint. Logging policies and sustainability are also discussed, but mainly in terms of production and consumption at a national or continental level. This analysis is extended on pp.250–254 in order to assess the impacts on the forest in more detail. Certain environmental measures of the 'quality' of the remaining forest are inferred, whilst on pp.254–257 there is an assessment of the interaction between agriculture and forest area, in particular the relative effects of subsistence and commercial farming. Finally, on pp.257–258, scenarios are used not so much for policy analysis but more for investigating the sensitivity of the model to assumptions about rural-urban migration.

In this analysis, Scenario D is rather different from the others. It is the only one that contains no explicit logging or environmental policy change. It was discovered that, although there might in theory be second-order effects, in practice these were very small. Simulation models produce such a plethora of output that it is sometimes difficult to comprehend. Therefore, analysis of Scenario D has been omitted from those sections that concentrate on the environmental impacts. The reader may infer that the results of Scenario D are very similar to those of the Base Scenario.

The Base Scenario

Before a detailed comparison of the scenarios is possible, it is useful to have a description of the main outcomes of the Base Scenario. The most fundamental assumptions are those concerning population and GDP and these are shown in Figures 8.1a–e. The results shown here concentrate on the regions producing tropical hardwoods and the main consuming regions (the advanced industrialized countries plus, because of its very large population, the rest of Asia). For definition of the abbreviated region names, see Table 7.1.

These projections do not warrant a great deal of comment. It is, however, worth noting the exceptional situation of Africa: its population growth rate is higher than any other region and, partially in consequence, its income per capita is falling while that of other regions is rising.

Having looked at the general situation, we are particularly interested in the output from the TROPFORM part of the model; that is the

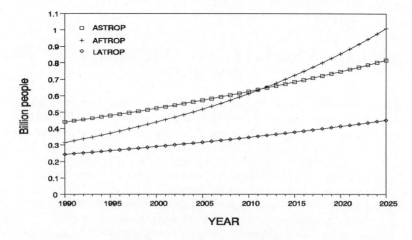

Figure 8.1a Populations – Base Scenario

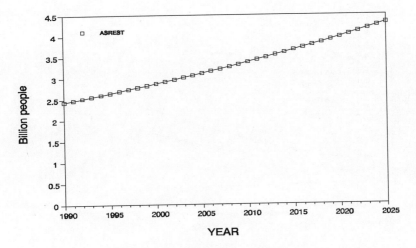

Figure 8.1b Populations – Base Scenario

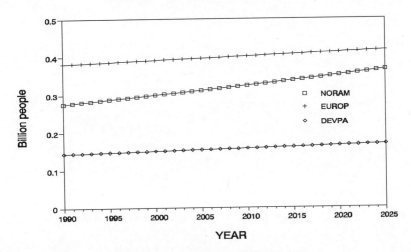

Figure 8.1c Populations – Base Scenario

information relating to forestry and wood production in the tropical regions. The most important indicators are presented in Figures 8.1f-j: logging activity, commercial reserves, exports, domestic consumption and total forest area.

In order to interpret these results, it is useful to review the way TROPFORM works. Demand worldwide for tropical hardwoods is driven by growth in population and income and is thus exogenous to the forestry sector. The model then assumes that there is an efficient market in these products so that not only is the market cleared, but global costs are also minimized. However, there are restrictions on this optimization process. There is a time horizon of twenty years built into logging operations; if in any country the current rate of timber

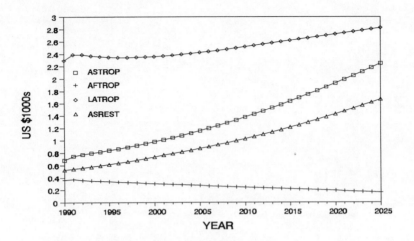

Figure 8.1d GDP per capita - Base Scenario

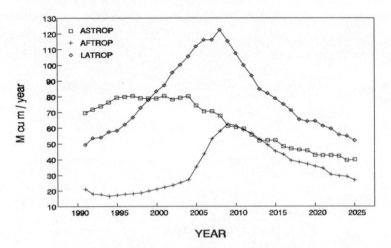

Figure 8.1e GDP per capita - Base Scenario

extraction would exhaust the reserves within this time horizon, then logging is restricted and, as far as possible, demand reallocated. One aspect of this reallocation is that domestic demand within the producing countries is given priority over exports. In addition, changes from one year to the next are kept within certain bounds to represent the realistic inertia of organizational change.

As can be seen from Figure 8.1g, reserves are falling and thus we can infer that the rate of logging exceeds that of regeneration. Ultimately, therefore, reserves would be exhausted and so the 'time-horizon' effect will come into play. Consequently, we would expect logging to grow with demand (as GDP and population grow) until the fall in reserves causes a reining in of logging activity. Indeed, this is the pattern of

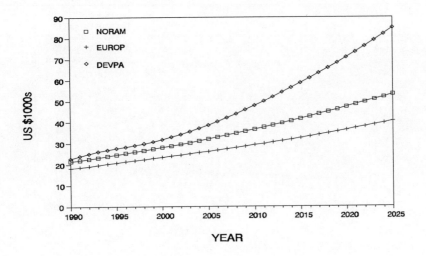

Figure 8.1f Logging activity – Base Scenario

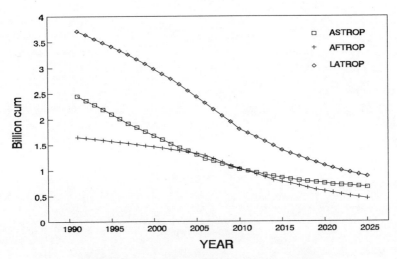

Figure 8.1g Commercial reserves – Base Scenario

exploitation that is observed with many finite resources, for example oil in the USA or coal in Great Britain (System Analysis Research Unit, 1977; Howe, 1987). We can see from Figure 8.1f that logging in Latin America and in Africa do clearly follow this pattern. The rise and fall is not so pronounced for Asia, the reason being that the twenty-year time-horizon effect starts to operate almost immediately, as opposed to around 2010 for the other two regions.

It is interesting to see how the production of timber is allocated between exports and domestic consumption. The 1993 LEEC report is

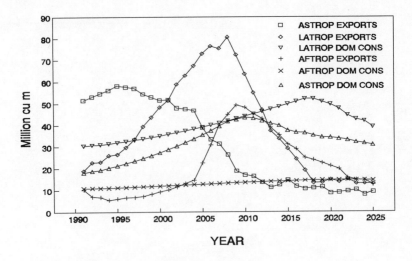

Figure 8.1h Exports and domestic consumption – Base Scenario

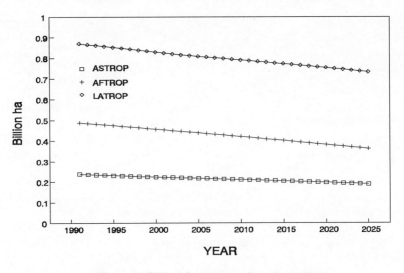

Figure 8.1i Total forest area – Base Scenario

clearly of the view that domestic use of hardwoods by the tropical producing regions will assume an ever greater importance compared with exports; Figure 8.1h very much bears out this conclusion. Exports from Asia start to fall soon after the start, those from Latin America and Africa peak around 2010 before falling thereafter. In contrast, domestic consumption rises in all producing regions until the loss of reserves causes a relatively slow decline from 2010 for Asia and around 2020 for Africa and Latin America.

Although commercial reserves are of great importance, it is also

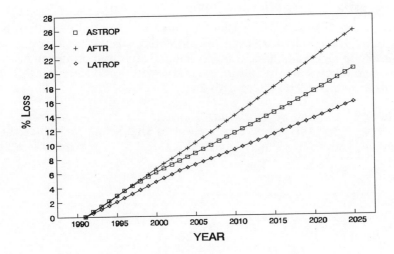

Figure 8.1j Total forest area – Base Scenario

necessary to look at what is happening to forests as a whole. TROPFORM is based on FAO statistics (FAO, 1989) which do not distinguish between different qualities of forest cover; in particular, there are not separate data on areas that have recently been cut for timber, and those which have been allowed some considerable time to regenerate. Logging per se does not, therefore, cause a loss of forest area; absolute losses in the model depend on agricultural activity (see pp.188–207). The total forest area remaining is shown in Figure 8.1i where the steady declines are clearly seen. However, the three regions start from greatly differing bases and so a percentage loss is, perhaps, more revealing; this can be seen in Figure 8.1j.

Tropical Africa loses the greatest proportion of forest, about a quarter by 2025, which is largely a reflection of its very high population growth rate. Although Latin America's absolute loss is the greatest (about 137 m ha), it experiences a lower percentage loss than the other regions because of its high initial stock. A further description of some of the land-use changes can be found on pp.254–257.

In summary, economic growth across the world causes a rising demand for tropical hardwoods. Asia's commercial reserves are the lowest compared with their production and thus practically no expansion of logging activity is possible and, indeed, output declines after 1995. Latin America, and later Africa, can expand their production to counter lower output from Asia, particularly in the export markets. However, from about 2010 even these regions have to cut back on logging to preserve their timber stocks. It is exports that experience the greatest rises and falls; domestic demand is much more steady, largely showing slight rises. The absolute loss of forest area, rather than changing quality due to logging, is much more driven by food production considerations. Thus the greatest fractional loss is observed in Africa where population will almost treble by 2025.

Comparison of the policy scenarios

The policy scenarios A-G described earlier have been implemented in the model. The results for Scenarios A-E which analyse policy options are shown in Figures 8.2a–u. A more detailed appraisal of the land-area issues then follows in which is included the effects of different levels of urbanization modelled in Scenarios F and G.

Scenario A. Sustainable forest management: tax transfer financed

The very important policy implication of this scenario is that the producing regions implement sustainable forest management practices

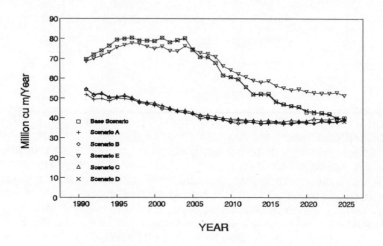

Figure 8.2a Tropical Asia – logging activity

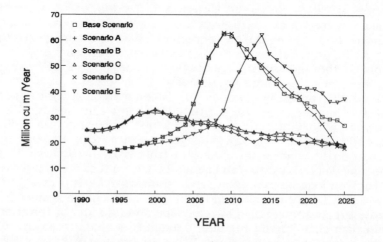

Figure 8.2b Tropical Africa – logging activity

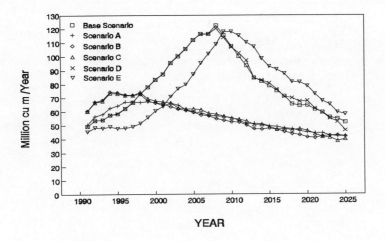

Figure 8.2c Tropical Latin America – logging activity

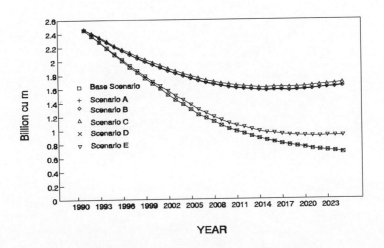

Figure 8.2d Tropical Asia – commercial reserves

as soon as possible. Their time horizon does, therefore, expand immediately to 40 years, although actual changes to planting policy take longer to effect. The longer term view considered on its own would tend to reduce production as operators become more aware of depleting reserves. In contrast, the imposition of an import tax in the advanced industrialized regions (North America, Europe and Developed Pacific) does reduce demand, partially mitigating the pressure on reserves (Figure 8.2j). Tropical Asia, which is already most affected by shortages of reserves, has to take immediate action in response to looking ahead 40 rather than 20 years. This effect is clearly

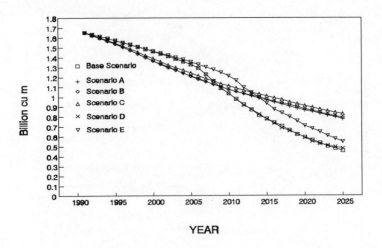

Figure 8.2e Tropical Africa - commercial reserves

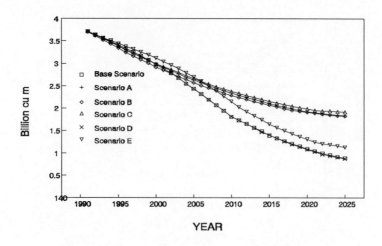

Figure 8.2f Tropical Latin America - commercial reserves

shown in Figure 8.2a where logging activity falls considerably (by over 10m m³ per year).

Figures 8.2b and 8.2c show something that may be slightly surprising at first sight; despite reduced demand and a longer time horizon, this scenario leads to an increase in logging in the first years of the simulation in Africa and Latin America. However, these rises are more than offset by the fall in logging in Asia. What has happened is that the longer view into the future has further emphasized the fact that Asia is using up its reserves at the greatest rate, and therefore some production has to be reallocated to the other two regions where the

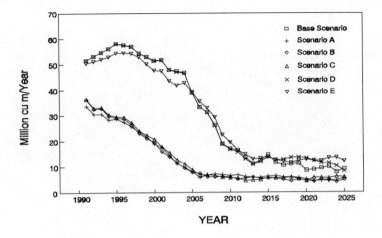

Figure 8.2g Tropical Asia - exports of tropical hardwood

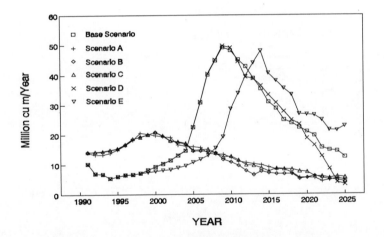

Figure 8.2h Tropical Africa - exports of tropical hardwood

pressures on reserves, although appreciable, are less. This phenomenon is relatively short term. The long time horizon soon leads to restrictions on production in Africa and Latin America too. The large rises in these two regions that occurred in the base run are no longer possible. Indeed, in all three producing regions, logging activity shows a gentle fall for most of the simulation period (Figures 8.2a-c).

As would be expected, the changes in logging, together with steadily improving regeneration policies, have important effects on commercial reserves (Figures 8.2d-f). The initial switch from Asia to Africa and Latin America is also apparent. However, by 2025 the reserves in all

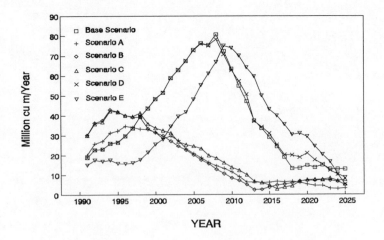

Figure 8.2i Tropical Latin America – exports of tropical hardwood

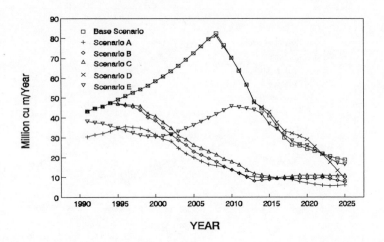

Figure 8.2j Advanced industrialized regions – demand for tropical hardwood

regions are considerably higher compared with the base case, being around double.

Figures 8.2g–j show what happens to trade. The immediate fall in imports due to the tax can be seen (especially in comparison with Scenario B which embodies the same policies on forest management without the tax effect). However, this is a relatively small effect in comparison with the changes to the very large peak in trade around 2010. Such a high level of exports is no longer possible; it would use reserves too quickly. It must be noted, however, that this restraint on exports is due to sustainable management policies which do not allow

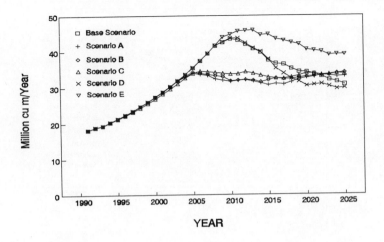

Figure 8.2k Tropical Asia – domestic consumption hardwood

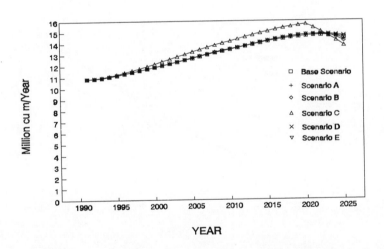

Figure 8.2l Tropical Africa – domestic consumption hardwood

over-exploitation of forests rather than the demand reduction caused by the tax increase. This conclusion arises from the fact that during the period 2000–2020 when the cut back in exports occurs, there is very little difference between Scenarios A and B. Figures 8.2k–m show that although sustainable management also reduces domestic consumption in Asia and Latin America, the effects are seen some ten years later and are much smaller in size.

Although this policy has been successful in conserving reserves, we must consider both the benefits and the costs. The tax does raise an average of $1500m per year over the period from 1990 to 2025. This is

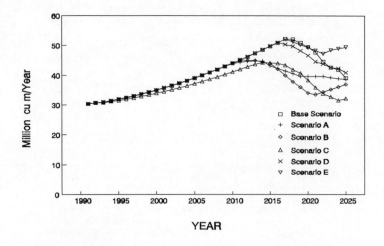

Figure 8.2m Tropical Latin America – domestic consumption hardwood

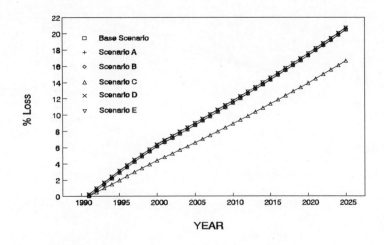

Figure 8.2n Tropical Asia – loss of forest area since 1991 (%)

achieved by levying the 40 per cent tax rate on an average figure for annual imports by the AICs of $3773m (compared with a figure of $9824m in the base run). This tax transfer is invested in the forestry industries of the producing countries. Also on the positive side, reserves are much greater and hope is held out for the survival of a long-term sustainable industry. However, there is a loss in the value of production and, probably more importantly, a loss of foreign currency earnings. These are detailed in Table 8.3.

One interesting inference from this table is the importance of trade in tropical hardwoods between the less developed countries, an

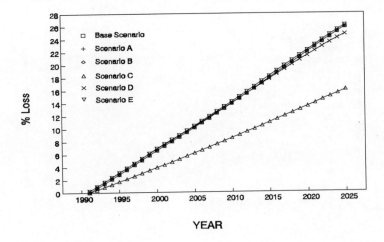

Figure 8.2o Tropical Africa – percent loss of forest area since 1991

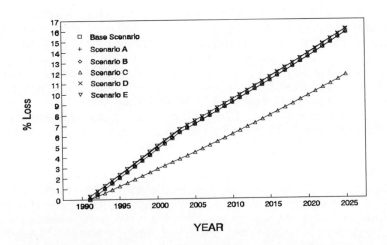

Figure 8.2p Tropical Latin America – percent loss of forest area since 1991

importance which grows yet further when exports to the AICs are reduced by the imposition of a tax. More information on this point can be inferred from Figure 8.2q.

Although these figures show cumulative losses that seem large in comparison with the gains in reserves, it is vitally important to remember that much has been omitted. The stock value of the commercial reserves is purely based on the timber value and there are many non-timber products that are not currently exploited (rattan, oils, resins, fruits etc.). In addition there are non-market benefits (e.g. watershed protection) and benefits that extend beyond the country

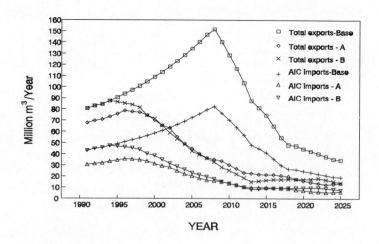

Figure 8.2q Total exports and AIC imports – Base + scenarios A & B

itself such as retention of biodiversity and climate control. A full assessment would include these non-timber and non-market valuations (Pearce, Barbier and Markandya, 1990, Chapter 5). However, the relevant decision makers in the tropical producing countries are working under simple accounting conventions, and if they are to forgo valuable production they will need to be convinced of the less direct benefits, and possibly also compensated for their losses.

On a final note, Figures 8.2n–p show that, because these scenarios have very little direct impact on agriculture, the absolute loss of forest is essentially the same as in the Base Scenario.

Scenario B. Sustainable forest management: aid financed

Much of what has been said about Scenario A also holds for Scenario B. The same sustainable-management policy is followed, the only difference being the way in which it is financed. No transfer tax is imposed; the $1500m per year is deemed to be given as aid. Figure 8.2q shows how, initially, the absence of a tax allows higher exports to the AICs. However by 2005 there is very little difference between the two scenarios as far as exports are concerned. The sustainable-management policy with its long time horizon has cut available supply so that the extra potential demand due to the removal of the tax cannot be met. Figures 8.2c and 8.2i (both relating to Latin America) are the only graphs of logging activity and exports showing any discernible difference between Scenarios A and B; one can thus infer that the extra exports to the AICs are being met by additional Latin American production.

The costs associated with this scenario are shown in Table 8.4.

Again, these figures are very much as for Scenario A with, as expected, slight increases in exports. The exports to the AICs have

Table 8.3 Scenario results: Scenario A

Values ($ billion)	Annual averages 1991–2025		
	Base Scenario	Scenario A	Difference
Logging	35.3	24.5	–10.8
Total hardwood exports	18.1	8.4	–9.7
Hardwood exports to AICs	9.8	3.8	–6.0
Tax transfer	0.0	1.5	1.5
Commercial reserves in 2025 (stock value)	402.2	850.0	447.8

increased more than the total exports which indicates that the tax-induced reduction has allowed the regions which are neither in the AIC or tropical-producer groupings (non-tropical developing countries and the former centrally planned economies) to increase their imports. Furthermore, the increase in logging is even less than the increase in total exports. The inference can therefore be made that imposing a tax and reducing AIC imports redirects production to other regions and to domestic consumption, making little difference to the total logging activity.

Scenario C. Sustainable development: aid financed
This scenario represents a very wide-ranging set of policy options which are designed to bring about 'sustainable development'. Although there must always be some question as to what 'sustainable' means when the world economy and population are still growing quite rapidly, the policies described by the World Bank as used in this scenario would certainly bring about considerable improvements to the natural environment of the developing world.

In addition to the sustainable forestry practised in Scenarios A and B, this scenario attacks two other key problems of sustainability, population growth and agricultural productivity. It is assumed, perhaps optimistically, that the additional investment required ($32 billion per year) is met by aid donations. This is an expensive programme, but the simulation will allow us to see what extra benefits are obtained in return for the extra investment.

Table 8.4 Scenario results: Scenario B

Values ($ billion)	Annual averages 1991–2025		
	Base Scenario	Scenario B	Difference
Logging	35.3	24.6	–10.7
Total hardwood exports	18.1	8.6	–9.5
Hardwood exports to AICs	9.8	4.6	–5.2
Aid transfer	0.0	1.5	+1.5
Commercial reserves in 2025 (stock value)	402.2	847.6	+445.4

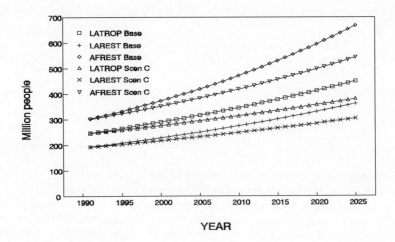

Figure 8.2r Populations – Base and Scenario C

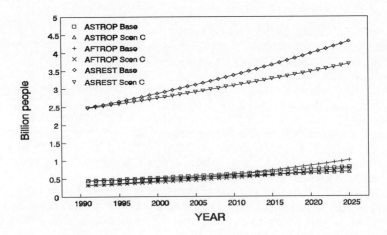

Figure 8.2s Populations – Base and Scenario C

As well as the direct environmental benefits, there is the reduction in population which is shown in Figures 8.2r–s.

Scenario C, with its very large aid flows, is the only scenario where GDP per capita differs appreciably from the Base Scenario. Figure 8.2t shows the changes for the three tropical regions, but it should be noted that all less developed regions benefit; most of the aid is not tied to forest conservation, but goes for more general development schemes such as agricultural soil conservation and female education. The 'Rest of Africa', already receiving the largest share of aid, sees its 2025 GDP per capita rise from $486 to $785. Tropical Africa also experiences a

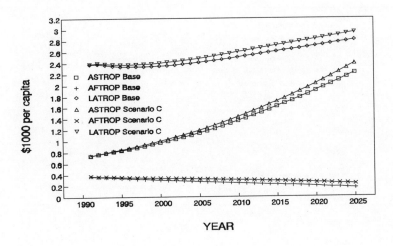

Figure 8.2t GDP per capita – Base and Scenario C

large increase in GDP per capita, from $159 to $229. What is also worthy of comment is that when this large amount of aid is given by the AICs, their GDPs per capita are not adversely affected; they too are higher by 2025, but only by a very small amount compared with the Base Scenario. The mechanisms within SARUM that generate these figures are quite complex and so it is not straightforward to give a clear causal explanation. Nevertheless, it does appear that the great stimulus given to the less developed economies does have subsequent benefits for the donor countries.

This scenario does have very much the same effects on commercial logging as Scenarios A and B, a not unexpected result given that the same moves to sustainable management are made. Any changes that are observable are beneficial; for example, the commercial reserves in 2025 are larger in this run than in any other. To see the important additional benefits of this scenario, we have to turn to an examination of agriculture and forest area. As Figures 8.2o-q show, this is the one set of policies that makes any appreciable difference to the total area of forest. Losses in Africa are almost halved, and by 2025 the world total tropical forest area is 97 million ha larger than in the Base Scenario. To put this figure in perspective, the World Resources Institute's (1990) estimate of annual deforestation was 12 million ha per year for the 1980s. The saving of 96 million ha thus represents eight years of deforestation at recent rates which are themselves the greatest ever experienced in history. From another point of view, this area is the about the same size as a country such as Tanzania or Nigeria. These policies, therefore, do make a very significant contribution to reducing deforestation, albeit at a high financial price.

The fundamental reason for the increased forest area is that less area is needed for food production. This in itself arises from two separate

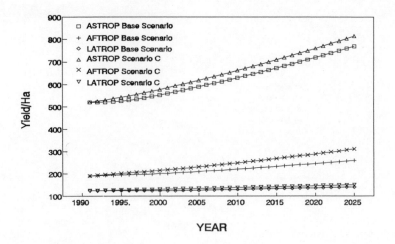

Figure 8.2u Yields per hectare – Base and Scenario C

effects: a reduction in population and a rise in agricultural yields. The population changes have already been looked at; the changes in yields are shown in Figure 8.2u.

It has to be remembered that the yield improvements shown above only apply to commercial agriculture; the subsistence sector is unaffected. However, the commercial sector is much larger and calculations based on their relative land areas and yield enhancements suggest that lower populations and higher yields each contribute about 50 per cent of the increase in forest areas.

Finally we need to consider the costs and benefits of these policies and these are shown in a similar form as before in Table 8.5.

The tropical producing countries are still losing a considerable value of timber production and exports, though not quite as much as in Scenarios A and B. The reduced deforestation does seem to have provided some extra resources. Although there are problems of financial incentives for logging companies, the countries as a whole actually experience a net improvement in terms of foreign currency;

Table 8.5 Scenario results: Scenario C

Values ($ billion)	Annual averages 1991–2025		
	Base Scenario	Scenario C	Difference
Logging	35.3	25.3	–10.0
Total hardwood exports	18.1	9.2	–8.9
Hardwood exports to AICs	9.8	5.0	–4.8
Aid transfer (in 2000)	0.0	32.0	+32.0
Commercial reserves in 2025 (stock value)	402.2	883.8	+481.6

the additional aid receipts are three times as great as the lost exports. When the increased GDP per capita is taken into account, it is clear that the tropical timber producing countries have not lost out financially at the macro level. Nevertheless, there will be a considerable restructuring within the countries. Many changes will, however, be beneficial such as improved education and agriculture.

Scenario D. Third World industrialization: trade induced

This scenario has no explicit environmental component, but the way in which it might have an impact is discussed earlier in this chapter. As would be expected, the GDP per capita rises slightly in those regions benefitting from improved market access for their exports, and falls slightly for the AICs who are purchasing more imports from the South. However, the differences are very small, all under 1 per cent in 2025 apart from the 'Rest of Africa' which is 4.5 per cent better off. The changes from the base scenario are shown in Table 8.6. What differences there are result in slightly less impact on forest resources.

Table 8.6 Scenario results: Scenario D

	Annual averages 1991–2025		
Values ($ billion)	*Base Scenario*	*Scenario D*	*Difference*
Logging	35.3	35.2	−0.1
Total hardwood exports	18.1	18.1	0.0
Hardwood exports to AICs	9.8	9.8	0.0
International transfers	0.0	0.0	0.0
Commercial reserves in 2025 (stock value)	402.2	406.1	+3.9

In summary we may conclude that global trade policies that do not have any direct connection with forestry do not have important secondary effects on tropical forests. More direct interventions, either in the markets for wood and wood products, or in forest management, or in agriculture are needed.

Scenario E. Change in consumer preferences

In this case there is a direct influence on the market for tropical hardwood products. Since the impetus comes from individual consumers, there are no government interventions at the macroeconomic level; only the markets respond. As would be expected, there is an initial fall in logging activities and, more markedly (see Figure 8.2j), in exports to the AICs. There is relatively little impact in Asian production and exports, but the effects are more clearly seen in the exports from Africa and Latin America. Essentially what this reduction in consumer demand does is to delay the very high peaks in exports by about five years. However, when we look at the exports to the AICs and the total exports, we see that the total exports still peak at about the same levels (though five years later) whereas the imports by the AICs attain a peak some 40 million cu m per year lower than in the Base Scenario. The exports have simply been diverted from the AICs to the rest of the world.

The well-intentioned consumers of North America, Europe and the Developed Pacific have cut their demands, but by the first decade of the next century, the tropical hardwood market is supply limited. Changes in pattern of demand do not, therefore, affect the total supply. Consequently if one group of customers cut their demands, someone else takes up their quota in this market where excess demand is chasing a limited supply. It does appear that more targeted, coordinated policies are needed to effect major changes. Nonetheless, although the consumers' actions have had little more than a redistributional and a delaying effect, the fall in commercial reserves is not as great as in any other of the policies that do not contain explicit reference to sustainable management. Furthermore, as will be discussed below, an improvement in forest quality also results.

The costs are presented in Table 8.7. The improvements are clear, but relatively small compared with Scenarios A, B and C.

Table 8.7 Scenario results: Scenario E

Values ($ billion)	Annual averages 1991–2025		
	Base Scenario	Scenario E	Difference
Logging	35.3	35.0	−0.3
Total hardwood exports	18.1	17.1	−1.0
Hardwood exports to AICs	9.8	6.7	−3.1
International transfers	0.0	0.0	0.0
Commercial reserves in 2025 (stock value)	402.2	518.3	116.1

Changes in forest quality

The figures produced by the model for forest areas simply give the number of hectares designated as forest, without attempting to measure the quality of that forest. For example, virgin forest, secondary forest and recently cut forest not yet allocated to any other purpose could all be included. The following graphs are thus included to give some indication of the quality of the forest remaining. Figures 8.3a–c show the average volume of commercial reserves per hectare of total forest; the more recent the cut, and the fewer trees remaining, the lower this figure will be. Figures 8.3d–f demonstrate another measure of how the forest is being used, the percentage of the total forest area currently being logged.

The base scenario gives a picture of increasing pressures from logging. The reserves fall to very low values in Africa and Latin America and, although the figure is higher in Asia, there is still a very marked decrease. The logging intensity of the forest rises correspondingly, an ever greater fraction of its area being under active logging. The association between high logging and low reserves for Latin America is

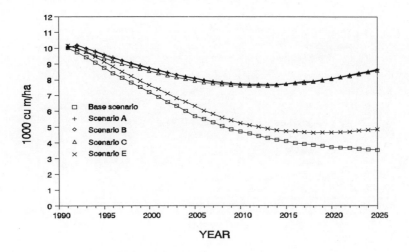

Figure 8.3a Tropical Asia – forest reserves/hectare

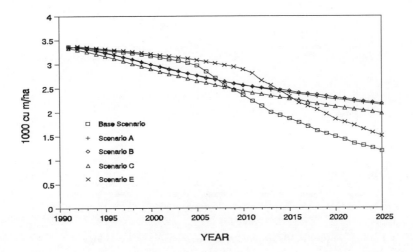

Figure 8.3b Tropical Africa – forest reserves/hectare

clear. It may also be noticed that, although these trends have much the same directions in all regions, they occur at different absolute levels. This results from differing endowments of the native forests in commercial species of tree.

Scenarios A and B show a clear improvement, as would be expected since they both incorporate the introduction of sustainable logging practices. The changes are very much the same for both scenarios, a consequence of the fact that both involve the same changes to forest-

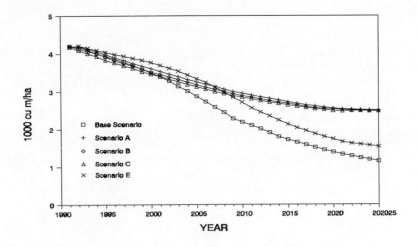

Figure 8.3c Tropical Latin America – forest reserves/hectare

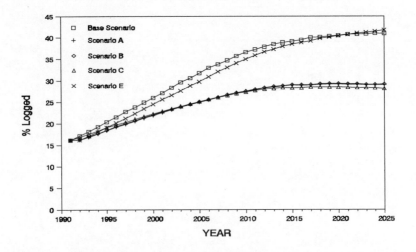

Figure 8.3d Tropical Asia – percentage of forest logged

management policies, only the method of finance being different.

The most noticeable improvement is in the reserves of Asia where the decline is reversed after about 2010. The restrictions on logging brought about by the new policy occur earlier in this region than in the others and thus this part of the world is the first to benefit. By the end of the simulation, the reserves per hectare are rising, and the percentage of the forest logged has reached a plateau. It is reasonable to infer, therefore, that the quality of the remaining Asian forests is improving as a result of these policy options. Although we do not see the same improvements as in Asia, there is evidence that the

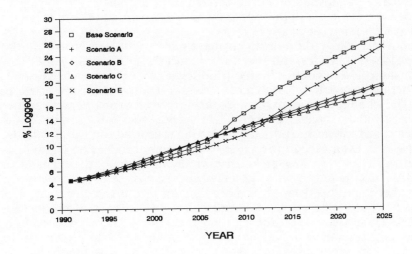

Figure 8.3e Tropical Africa – percentage of forest logged

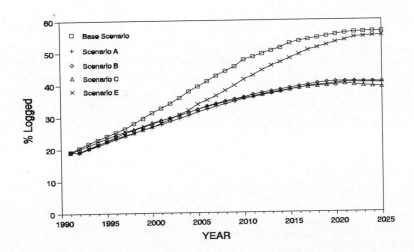

Figure 8.3f Tropical Latin America – percentage of forest logged

degradation of Latin American forests is levelling off. However, even such relative improvement does not occur in Africa where the percentage of area logged continues to rise with a corresponding fall in reserves per hectare. Such a result is again due to the timing of the restrictions in logging brought about by these policies. As was mentioned earlier, the initial restrictions in the output from Asia lead to a substitution of the supply from Latin America, which, in turn, starts to restrict its supply with Africa making up the shortfall.

Scenario C also includes the same forest management policies as Scenarios A and B, and so the pattern of changes is much the same. However, the lower population growth does have a significant effect on demand, seen most clearly in the reserves per hectare on Latin America and Africa. Since there is no change to the population growth in the developed countries, this result reinforces the point made earlier that consumption of timber by the producing countries themselves (and other non-tropical developing countries) is an important consideration.

The changes in Scenario E do bring about improvements to the condition of forests, especially in Africa and Latin America. As mentioned earlier, the peak in production is delayed compared with the base run, even though the value of maximum production is much as before. The delay does mean that, at any given time, the cumulative production is lower, with the consequence that reserves are higher. Furthermore, this delay gives more time for forests to regenerate, thus further increasing reserves. Because of the succession of production peaks from Asia to Latin America to Africa, these last two regions have the longest time to benefit from regeneration and thus gain most. In conclusion, while changes in consumer preference have relatively little effect at the macroeconomic level, there are appreciable benefits to be gained in terms of forest quality; less is logged and there is more time for regeneration to occur.

Changes in land use: commercial and subsistence agriculture

The model produces a large amount of information on land use, some of which was shown in Figures 8.1i and j earlier. We are interested in the loss of forest area, and particularly the causes of that loss. The change in quality due to logging practices has already been discussed but, in this model, irreversible loss of forest is due to the expansion of agriculture. Figure 8.4a shows how land use changes over time in the

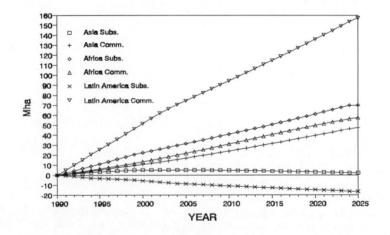

Figure 8.4a Cumulative loss of forest to farming – Base Scenario

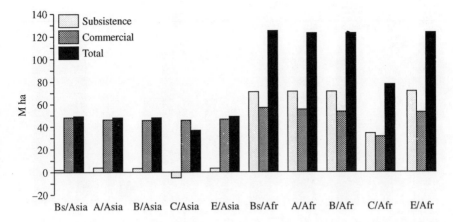

Figure 8.4b Cumulative losses to agriculture – To the year 2025 by Scenario and region

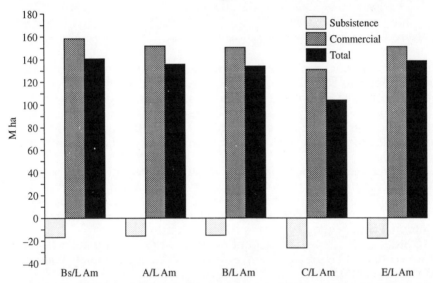

Figure 8.4c Cumulative losses to agriculture – To the year 2025 by Scenario and region

Base Scenario. The expansion of commercial agriculture across the world is very clear, and is the inevitable consequence of a growing population (with an ever increasing fraction in urban areas) and an increased standard of living. The effects of the regional variation in urbanization are also apparent. After a small rise up to the year 2000, the migration from rural areas in Tropical Asia starts to outweigh the growth in total population, and the area required by subsistence farmers starts to fall. Indeed this decrease indicates that farmers are leaving the forest areas, although the amount of released land is small compared with the extra required by commercial agriculture. Tropical Latin America shows a similar change, except that it commences from

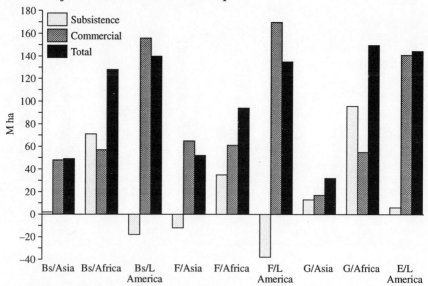

Figure 8.4d Cumulative losses to agriculture – To the year 2025 by Scenario and region

the beginning of the simulation; the projected rate of urbanization is great enough to reduce the land required by the subsistence sector. Africa, with a lower level of industrial development compared with the other two regions, continues to have a growing land requirement for subsistence agriculture right up to 2025. Indeed, the amount of forest land needed for this purpose is greater than that needed for commercial agriculture.

The three regions, therefore, fall into different categories in terms of the land requirements of subsistence agriculture. Latin America has a falling requirement; in Asia there is a small rise followed by a gentle fall; and in Africa there is a continuing steady rise which even outstrips the additional land needed by commercial agriculture.

In order to condense the amount of information generated by all the other scenarios, Figures 8.4b and 8.4c only show the cumulative losses up to the year 2025.

How are these changes in land use affected by the different policy scenarios? Remembering that absolute changes to forestry areas are effectively controlled by changes to agricultural land, it is not surprising that there is very little difference between Scenarios A, B and E (Scenario D is not analysed here for the reasons given above). These all follow the same agricultural policies and therefore have little effect on the conversion of forest land to agricultural use. They do, however, make a difference to the condition of the forest as discussed earlier.

Scenario C incorporates major changes in agriculture and food consumption. The lower population coupled with the higher yields per hectare both move to reduce the land needed for food production. However, the change is relatively small in Asia compared with that in the other two regions. The reason is that Asia has the smallest agricultural area in relation to its food demand and thus has to follow more intensive farming methods, one manifestation of this being its greater use of fertilizers. This region is thus more affected by diminishing marginal returns to fertilizer, with the corollary that the marginal cost of fertilizer is higher. When demand is reduced, as in Scenario C, rather than using less land, it is more economic to reduce fertilizer use because of its high marginal cost. This effect is confirmed by the fact that, in 2025, the fertilizer application in Scenario C is 86 kg/ha compared with 243 kg/ha in the Base Scenario.

Subsistence agriculture is determined by population size alone, there being no possibility of substitution between intensive and extensive techniques. Thus the assumptions about the lower growth of the population built into Scenario C result in a direct reduction of the land needed for subsistence agriculture. Such changes are apparent for all regions. Asia sees a reversal of the amount of land taken from the forests, with a net reduction by the end of the simulation. The land taken from the forests in Africa is lower in 2025 by some 30 million hectares, whereas in Latin America, the cumulative amount of land released by subsistence farmers is lower at all times.

The effects of urbanization

Finally, we move from analysing the effects of different policies to investigating the sensitivity of the model to certain assumptions, those concerning the rate of urbanization. Nevertheless, it should be remembered that social and economic policy can have an effect on migration and thus this section is not without its relevance to policy issues.

Figure 8.4d shows the losses of forest up to 2025 for Scenarios F and G in comparison with the Base Scenario. It should be remembered that what we are investigating here is the sensitivity of the model to changes in urbanization rather than an analysis of a particular policy. Given the uncertainty in predicting migration, variations of ± 25 per cent in the non-urban population in the year 2025 would not seem an excessive margin of error. We see a similar pattern for Africa and Latin America. Increasing urbanization (Scenario F) reduces the land lost to subsistence farming and increases that lost to commercial farming. Indeed, the greater urbanization actually results in a decrease of the forest area required for subsistence agriculture. However, over the 35 years of the simulation, there is a large enough increase in the area required for commercial agriculture to more than offset this decrease; there is always a net loss of forest. The reason is very straightforward. Commercial agriculture has higher yields and thus, strictly in land-use terms, it is more efficient to feed a city dweller than it is for someone to feed themselves by subsistence farming. However, this result does depend on the balance of yields between commercial and subsistence agriculture. The data used for Asia do not demonstrate the same effect. The relatively higher yield for subsistence agriculture in this region, together with its relatively high urban population already, means that the total loss of forest area is relatively insensitive to changes in urbanization.

Scenario G, in which the 2025 rural population is 25 per cent higher, shows very similar effects, but in the opposite direction. In Africa and Latin America, the rise in land area required for subsistence is larger than the fall in commercial agricultural land; on balance, more forest is lost.

In conclusion, there certainly seems prima facie evidence that the amount of forest lost is affected by changes in urbanization. Moreover, the scale of the effect may well be quite large. In Africa, for example, the difference in total forest area between Scenarios F and G is some 50 m ha. Such figures must, however, be treated with caution. Accurate forecasts of migration are very difficult to produce and all we have done is to specify a range of possible values that intuitively appear not implausible. Furthermore, in a global model such as this, it is not feasible to model all aspects in great detail; subsistence agriculture land-use is one area that would benefit from a closer study dedicated to this particular subject.

CONCLUSIONS

A range of different policies have been analysed which cover different approaches to forest conservation and development issues. We can first conclude, by looking at Scenario D, that policies with no specific 'environmental' content have little effect on forest resources, except in so far as they affect some of the economic driving forces such as GDP. If overall consumption is not much changed, then neither is the impact on the forests.

Of more effect is Scenario E in which consumer preferences change. However, this brings into play a mode of response that is important for all our analyses using the model. The market in tropical timber becomes supply-constrained around the turn of the century. In effect, consumers are being rationed because reserves are inadequate to meet world demand. Thus if any one consumer reduces their consumption, this allows another customer's rationing to be eased. The net effect is a reallocation of demand, but relatively little change to production and use of resources. Hence, if consumers in the advanced industrialized world voluntarily lower their demand, exports to other regions rise as does domestic consumption within the producing regions. If consumers are to reduce their impact on the forests, the change in sentiment really needs to be worldwide. Nevertheless, the delays in the exploitation of timber reserves do have an unequivocally good effect on the quality of the forest.

The mechanism of supply constraint also has an effect on the global pattern of timber extraction. There is a succession of production peaks: Asia, then Latin America and finally Africa. The absolute timing of these peaks, and their actual size, can obviously be affected by different policies, but the broad pattern remains the same. Clearly, if a conservation policy was initiated in only one of these regions, then some change to this ordering might occur, but simple changes to global demand leave the pattern of succession unchanged. Such results do have implications about where the pressures lie and where the most urgent action is needed. However, the time available is short; in the Base Scenario even Africa is only fifteen years away from its production peak.

A more coordinated approach to conservation is for the tropical countries to move towards sustainable forest management. The cost of this is considerable, probably at least $1500m per year. However, if this policy is financed, there is a marked improvement in commercial reserves of timber, but this is at the expense of lost production and exports. Further research is definitely needed in the financial analysis of sustainable forestry. The *prima facie* figures presented here show the annual losses considerably outweighing the additional funding that comes from aid tax transfers. Against this must be set the higher future reserves and other potential sources of value within the forests, but on balance it appears that it will not be easy to convince the producing countries that a policy concentrating on sustainable forestry would be financially viable.

With regard to a sustainable forestry policy, the two ways of financing it have been analysed. The main difference between them is that a tax in the AICs reduces demand via the price elasticity. In character this is similar to a change in consumer preference – with the same outcome. The reduction in AIC demand has some effect on reserves, but much of the effect is lost as other consumers step into the market to take up the shortfall in demand. The tax option, therefore, has limited advantages. However, as in Scenario E, there is an improvement in forest quality.

One can also look at a much larger sustainable development aid programme. Although this would cost $32 billion per year, provided it has the planned impacts, it clearly yields the greatest environmental benefits. Not only are commercial timber reserves improved through sustainable management, but the benefits to the tropical regions appear comparable with the costs. In terms of foreign currency, the aid inflows exceed the timber-export losses, but the question does remain as to how the logging companies within the tropical producing countries will be compensated for their lost production. However, this policy has a very beneficial effect on forest conservation through a reduction in the land needed for agriculture; the lower population and the higher yields both contribute. By 2025, the world has almost 100 million ha more of tropical forest.

The most difficult question to answer in relation to this package of policies must be 'who pays?' An extra $32 billion represents under 0.2 per cent of the industrialized countries' GDPs but, in a recession, it is difficult to see this being forthcoming. Furthermore, the donors would have to be convinced that the aid was being well spent. If such problems can be overcome, there is a clear hope of considerably reducing deforestation, but it will require an action programme that tackles all development issues, and does not approach deforestation in an isolated manner.

The model has also allowed us to make some inferences about environmental impacts on the forests. It is clear that large-scale losses of forest are primarily due to the expansion of agriculture. Admittedly logging activity can change the nature of a forest: unless it is then converted to some other use, some sort of forest will regenerate. Even in the worst case (Asia in the Base Scenario) no more than 45 per cent of the total forest is actively logged. The greatest absolute losses are due to commercial agriculture in Latin America, followed by subsistence agriculture in Africa. There are various driving forces causing such changes. Most simply, however, it is a growth in demand for food. Growth in population is probably the most important cause of this rise in food demand, but it must be remembered that an increasing standard of living also has an effect, particularly as it is associated with a rise in demand for animal products which have a much lower conversion efficiency and thus require more land for a given final calorific value. In addition, without getting into the complex debate about the ethical issues of population policy, we can infer that lower population growth does mean that less forest will be converted to agricultural use.

Rural–urban migration affects the balance between commercial and subsistence agriculture and is certainly an area where our understanding of the processes is somewhat limited. There is evidence that increased urbanization results in less loss of forest because food production is concentrated in the higher-yield commercial sector. However, such inferences do depend on many crucial assumptions. For example, might it not be possible to improve methods in the subsistence sector so that their yields rise too? What one can infer unequivocally is that improvements in agricultural practices throughout the societies of the tropics will have benefits for forest conservation.

In summary, our results show that there are many influences on the quantity and quality of tropical forests in this world. Continuance of present trends will result in a rapid diminution of commercial timber reserves: over half will be used by 2025. Similarly, the necessary growth of food production in the tropical forests will lead to a considerable loss of forests to agriculture. The many influences, and their complicated interactions, mean that there is no single policy that will effectively limit loss of timber reserves and forests. The limits on supply mean that if one group forgoes consumption of timber, some other group will take over the demand instead, although any delays in exploitation do have a good effect on the quality of the remaining forest, as well as allowing time for more effective policies to be developed. Any large-scale conservation of timber reserves requires action by the producing countries. Such action, however, has serious economic consequences in terms of lost production and, perhaps more importantly, lost export revenues. It is difficult to see how the producers will do this voluntarily without some form of recompense. Taxing imports by the advanced industrialized countries does not appear to raise enough revenue to cover losses in exports. It thus seems that some form of aid transfer is required. Although such conservation policies preserve timber reserves and thus improve the state of the forests, absolute loss of forest areas is mostly due to agricultural expansion. A comprehensive range of development policies which includes population programmes and improvements to agricultural productivity, as well as the implementation of sustainable logging practices, is the most effective way of conserving reserves and reducing forest loss. It is very expensive (around $32 billion per year) and thus it is difficult to see where the finance will come from. On the other hand, the transfers to the producing countries exceed the loss of earnings from timber, thus overcoming one fundamental objection. Policies that can conserve the forests are, therefore, not unfeasible. It is a challenge to policy makers and governments throughout the whole world to make them work.

APPENDIX 1

PUBLIC POLICY AND DEFORESTATION IN AMAZONIA

INTRODUCTION

In this chapter the economic developments and policy measures which contributed to the current deforestation problem in the Northern part of Brazil, Amazonia, will be outlined. It is concluded that much of this deforestation problem can be traced to the various regional policies that the Brazilian government has implemented in the course of time in its efforts to develop the agraricultural sector in Amazonia.

According to Mahar (1989) Brazil is covered with almost 3.5 million square kilometres of tropical forests, including transitional forests (a more conservative estimate is made by Myers (1991), who believes that the tropical forest only covers 2.2 million km² of the Brazilian territory). In any case, it is generally agreed that more than 30 per cent of all tropical forests in the entire world are situated in the 7 million km² large Amazon river basin (Anderson, 1990a). This basin stretches out not only over large parts of Brazil (approximately 5 million km²), but also over Bolivia, Peru, Equador, Columbia, and Venezuela. The 5 million square kilometres of legal Amazonia, as the Brazilian part of the basin has been called since 1960, cover almost 60 per cent of Brazil's land surface.

The tropical forests of Brazil became increasingly subject to human encroachment in the 60s, a process which accelerated in the 70s and 80s.

The first scientific airborne estimates of tropical deforestation, made in the early seventies by RADAM (Radar na Amazônia), indicated that still fairly little deforestation had taken place in the 60s. In 1975, however, Landsat satellite images suggested that by then already 2.8 million hectares, 0.6 per cent of Legal Amazonia, had been cleared.

The area cleared expanded from 7.7 million hectares in 1978 to 12.5 million hectares in 1980. By 1988, almost 12 per cent of the tropical forests in Legal Amazonia were already cleared, totalling some 60 million hectares (Mahar, 1989). Estimates of the average annual deforestation rate in the area in the 1980s ranged from a modest 0.9 per cent according to Schmidt (1990), and a moderate 1.8 per cent according to WRI (1990), to a pessimistic 2.3 per cent according to Myers. The yearly range of deforestation in the early 90s will probably lie somewhere within in a range of 1.7 and 8 million hectares per annum (WRI, 1990).

Although aggregate clearing rates do give an indication about the size of the problem in Legal Amazonia, they do not provide information about the spatial distribution of this rate over the different subregions. Deforestation rates in the various regions diverge for instance from 0.4 per cent in Amapá to more than 23 per cent in Mato Grosso and Rondônia (Mahar, 1989). This wide variety of, on average, high deforestation rates is related to the various patterns of regional economic development, in particular in the agricultural sector.

According to Mahar (1989), cattle ranching, annual cropping and logging are responsible for the largest part of the decline of the Amazonian tropical forests. Other causes of deforestation are the construction of infrastructure, for example roads and hydroelectric dams, mining, silviculture, and perennial cropping. All these factors are interrelated however, which makes them nearly imputable. For instance, large tracks of forest are sequentially logged, turned into fields, and finally, after three or four years, turned into pastures.

Recent estimates by Amelung and Diehl (1992) of the contribution of the various factors indicate that agriculture is responsible for 89 per cent of Amazonian deforestation (more precisely: pasture 40 per cent, arable land 32 per cent, shifting cultivation 13 per cent, and permanent crops 4 per cent). The remaining 11 per cent, according to this study, is caused by hydroelectric production (4 per cent), mining (less than 3 per cent), and forestry (2 per cent), mainly for charcoal production and an unexplained residual of 2 per cent.

The underlying reasons why Brazilians increased their farming activities in Legal Amazonia are:

- the favourable fiscal situation due to public policy,
- the prevailing poverty, partly due to the unequal land and income distribution in Brazil as a whole, and
- ecological ignorance.

A comprehensive analysis of deforestation in the Amazon region should therefore not be limited to the region itself but should also take the interregional relationships between Amazonia and the other parts of Brazil into account.

In contrast with the causes of deforestation in Africa and Asia, the growing population size does not seem to be a major explanatory factor in Brazil. According to Bilsborrow and Ogendo (1992), the average land area available to an economically active person in agriculture in Brazil in fact increased, from 2.4 hectares in 1965 to 5.7 hectares in 1987.

In the next section's attention will be given to the role of public policy in the deforestation process in Amazonia interacting with the other factors mentioned earlier. The Amazonian case serves to illustrate once again that deforestation processes often are determined by a combination of factors that should not be dealt with in isolation.

AMAZONIAN LAND OCCUPATION: AN ECONOMIC ANALYSIS

In analysing the main underlying causes of deforestation in the Amazonian area, we will, again, start from the assumption that much of the deforestation can be traced to the functioning of the market mechanism in conjunction, paradoxically enough, with the existence of several market failures. Over the next few pages a survey is given of market conditions that have provided the incentives for overexploiting the tropical forests in Amazonia. From an economic point of view, the main characteristics of the Amazonian situation are the abundancy of cheap land and the scarcity of labour and capital. According to neoclassical theory, profit maximization will take place if the marginal products of the inputs vary in proportion with their prices. Agricultural development in Amazonia will thus take place in a land intensive way, at least if the assumption of decreasing marginal productivity is made.

The above implies that since there is plenty of potential farmland, rational investors are commonly willing to buy a large area of land as soon as they start a ranch, and to produce in a land-using way. One should not overestimate the long-term agricultural potential of land in Amazonia, however, because the soil of the former rainforest is almost everywhere fairly infertile.

There are various reasons explaining why extensive farming and ranching in Amazonia offered a highly profitable investment.

First, the government offered large tax credits and subsidies to the cattle ranchers to lower their investment costs. Investments that would have been highly unprofitable without governmental support, turned out to be very profitable once they were subsidized with the help of public money.

Second, a major part of the fixed costs involved with the Amazonian agricultural production, i.e. the costs due to infrastructural facilities, are also fully borne by the same government. Moreover, the costs of past government investments in infrastructure in the region, such as road building for instance, to initiate migration into the jungle, have never been passed on to the population of the region, even if the maintenance costs are considerable in the moist tropics.

The main reason why the Brazilian government has tried to open up Amazonia for economic activity was to cope with Brazil's major 'poverty pollution', which was generally perceived as of much more importance than the preservation of the tropical forests. Therefore, extensive programmes for the development of agriculture in Amazonia were developed, even if Brazil had enough agricultural potential elsewhere. If all the land area would be evenly distributed over the total population, each person would have four hectares at its disposal. However, in reality, 4.5 per cent of the population presently owns some 80 per cent of the country's farmland; 70 per cent of rural households are landless (Anderson, 1990a).

Third, with respect to land use one notices an externality problem because private investors commonly consider the costs of environmental degradation as none of their concern as long as this degradation does not directly affect their production and income. However, when the impact of all investment projects in Legal Amazonia is taken into account, it is clear that their long-term ecological consequences through deforestation – and therefore their indirect consequences for production – are considerable for Brazil as a whole. There is strong evidence that the partial removal of the tropical forest can have dramatic consequences for South America's water cycle, which makes it a matter of speculation what the long-term perspectives for agricultural production in the area really are.

Fourth, the farmers' social discount rates are high. The cleared pastures and fields are only productive for a few years, because the soils loose their nutrients – which they mainly acquire after burning – fairly quickly. The social (opportunity) costs of deforestation, such as the loss of by-gone gathering earnings, last far longer. It takes years, if not decades, for a forest to regenerate, and to make forest exploitation possible again. The high discount rate illustrates some sort of defeatism, which is partly related to poverty.

In order to change a situation wherein large tracks of forest are converted into uneconomic farmlands, the Brazilian government could try to change the relative prices of land, capital and labour, for example by limiting the amount of land allotted to each farm. Another possibility would be to stimulate the research into land-saving techniques and the promotion of sustainable management.

One has to make a distinction between the substitution effect and income effect of subsidizing production factors, however. On the one hand subsidizing labour or capital alters the relative price of land and therefore in theory stimulates the farmers to substitute some of their land for the subsidized inputs. On the other hand – and this is what one often sees in practice – the other effect may dominate, according to which the overall farmers' demand for land increases due to the overall cost reduction.

Moreover, the steadily rising land prices in the recent past, which were higher (on average) than overall inflation (see Figure a.1), increased the demand for entitled land for speculation purposes. It is quite possible that this land speculation increased the price of land above its net present value of all the harvests to come. One of the most important indirect effects of this phenomenon is an increase in deforestation caused by squatting. On the one hand the poor landless farmers cannot afford to buy the expensive farmlands; on the other hand the landowners have a strong incentive to try to capture more land titles, and therefore to also turn to the tropical forests.

Only during the two year period after February 1986, when the government developed the Cruzado plan to eliminate (expected) inflation, was the rise in land prices less than the overall inflation. After the collapse of the Cruzado plan in the beginning of 1987, the rise of land and consumer prices accelerated spectacularly. By the end of 1987 investments in land purely for speculative purposes had become profitable again.

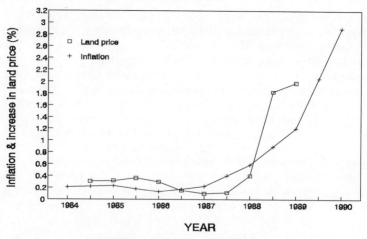

Figure a.1 Land prices and inflation

Sources: Anuario estatistico do Brasil (1990), and the Statistical Yearbook for Latin America and the Caribbean (1991).

Note: the land prices used in this figure are prices of the 'free lands' in Brazil, which means that they can only serve as an indication for the specific prices in Amazonia, and that therefore no conclusions for Amazonia can be drawn on the basis of this figure alone.

If the present-day land degrading effects of the agricultural practices in the region were fully taken into account in the land prices, the value of land would probably not rise at all. A sustainable agricultural production, e.g. by a more intensive use of chemical fertilizers, would in any case require a serious amount of additional resources. This would push future profits down, and thus the present land values.

In sum, the past phenomenon of ever increasing land prices in Amazonia remains difficult to explain without any special references to the incentives provided through government policies:

1) Investment in land is generally perceived as a fairly reliable shield against (rapid) inflation. This may partly explain why Brazilian investors generally have been rather keen to acquire land titles in Amazonia. This may have initiated a process of self-fullfilling prophesies with respect to increasing land prices and the perception of rather small risks.

2) Land in Amazonia serves as a tax-shelter, because the taxes on corporate agricultural profits are far less than the tax rates on other Brazilian corporate profits. This provides investors with an incentive to invest in agricultural land even though the pre-tax profits are considerably lower than in non-agricultural investments.

There is an important dynamic dimension to the tax shelter motive as well, i.e. as long as the tax system is not completely indexed. Because the effective capital gains tax rate for the possession of Amazonian land is less than the ordinary income rate in Brazil, the negative impact of inflation on the real yield on land will be less than its negative impact on the real yields on other financial assets in Brazil. In theory, the demand for land will therefore increase until an equilibrium is reached whereby the relative rise of the price of land has equalled the real yields of all the assets in the portfolio of an average Brazilian investor. This process that will lead to a rise of the relative price of land will continue as long as the expected rate of inflation is moving upward and/or indexation to eliminate those nominal gains has not been fully implemented in the tax system.

3) Expectations about possible increases in the availability of tax-credits and state subsidies press current land prices upward, because these future gains will be discounted in the current land value.

These three reasons alone cannot explain a situation of land values increasing into infinity, however. Only if the inflation rate accelerates and/or the government continues to increase its financial support to the land owners, are land values expected to rise. Those situations are not sustainable and are therefore likely to end in the (near) future. Until now, however, land values have continued to rise, making cattle ranching and small-scale farming to a large extent responsible for the current deforestation problem. These particular types of agricultural activities will therefore be elaborated further in the following sections.

The economics of cattle ranching in Amazonia

A typical farm in an Amazonian settlement project has an average size of 100 hectares of land and an average cattle ranch of at least 3000 hectares. In order to comply with the 50 per cent rule – a policy measure to relieve the pressure on the forests by forbidding farmers to use more than 50 per cent of their land as farmland – both types of settlements mentioned have to keep half of their land area as forest. The average rancher therefore can clear 1500 hectares of land for his herd, and a farmer 50 hectares for cultivation purposes. Although in practice it is fairly easy for both types of settlers to acquire the possession rights of such large tracks of forest, e.g. by slashing and burning it, it is almost impossible for them to keep the cleared parcels in productive use, at least without some additional investments in agricultural equipment and without the assistance of external labour.

The average rancher's demand for additional labourers (VMP in the left part of Figure a.2) remains limited, however, because the labourers' productivity on the infertile farmlands is low as compared to the regional reservation wage level. (The reservation wages are high in Amazonia, because in its open frontier situation, farmworkers have the opportunity to substitute wage income by gathering and hunting. Furthermore they can earn a living by subsistence production of foodstuffs.) The supply of labour will in any case be small at the offered wages. In some very infertile areas a non-zero equilibrium at the labour market may under the current conditions be nonexistent.

Similar problems arise with respect to capital procurement. Due to the high risks involved in agriculture, real interest rates are high. This effectively discourages the demand for capital. Just as was the case with the labour market, a non-zero equilibrium in the capital market for agricultural investments may well be nonexistent.

If a market equilibrium for commercial agriculture under these technical and economic conditions is not feasible at any positive wage rate and interest rate, from an investor's viewpoint it is irrational to become involved in this sector. The widespread existence of cattle ranching in the Amazon area indicates, however, that in practice (partial) solutions have been found to make ranching profitable. Such 'solutions' are:

- tying labourers to farms by extra-legal means. The ranchers offer loans to new immigrants, which normally cannot be repaid. The indebted persons are therefore forced to stay at the farm, a system known as 'aviamento';
- enclosing the frontier, or at least, denying labourers the access to alternative means of existence, by destroying the surrounding virgin forests; and
- providing public capital through subsidies and tax refunds, in order to lower interest rates on investments and raising the marginal productivity of labour.

The first two options shift the labour supply curve downwards. These options can, however, be counterproductive in the long run, because they encourage outmigration. The third option shifts the labour demand curve for the commercial agricultural sector upwards.

As stated earlier, the situation in the Amazon region cannot be regarded in isolation from developments in the rest of the country. Interest rates and the prices of agricultural products particularly are for a large part determined in the national markets. Legal Amazonia is a price taker both in the capital and agricultural output market, because

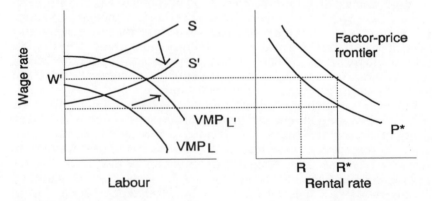

Figure a.2 The regional labour market   **Figure a.3** The production technology curve

Source: Kyle and Cunha, 1992.

of the marginal role of the area in the Brazilian economy (3 per cent of national income and 5 per cent of the agricultural production). In contrast with the former two markets, Amazonia is not a price taker in the labour market. The wage rates for agricultural work in Amazonia are higher than in other parts of Brazil. This implies that rural workers receive premium payments to compensate for their harsh living conditions in the tropical climate.

The Amazonian case can therefore be summarized as follows. Although it is possible to clear either the Amazonian labour market or the capital market, it is not possible to clear both of them simultaneously. At least, not as long as factor prices are high and output prices low. This point is illustrated by Figure a.3, wherein a factor price frontier (P*) is drawn for the current situation of high input prices and low output prices. The frontier unites all those combinations of factor prices which are consistent with a given output price, based on available technology.

A non-zero equilibrium is therefore only possible if the factor price frontier moves outward. Up till now, this was achieved by providing subsidies to the investors, thereby lowering real investment costs. With respect to Figure a.3 this means that capital is provided at a rate of R, instead of against the market rate, R*. The subsidy (in per cent) would therefore be (R* − R). When capital is subsidized one may expect this to have a positive effect on the demand for labour. That is, the scale effect is expected to outweigh the substitution effect.

Without the use of subsidies, the frontier can move outward if technological progress lowers input prices (which is unlikely as long as capital is scarce) and/or by an increase in the output prices, which are determined elsewhere outside the region. The theoretical analysis leads us to the conclusion that scarcity of labour and capital is only a symptom of the economic unfeasibility of agricultural projects undertaken in the Amazon (Kyle and Cunha, 1992).

A case study

Hecht, Norgaard, and Possio (in Mahar, 1989) analysed the internal rates of return of 'a typical ranch' in twelve different situations. They constructed a simulation model for a 20,000 hectare ranch receiving a 75 per cent tax credit. In each situation a distinction was made between corporate and total invested capital.

The results as presented in Table a.1 make it perfectly clear that, due to the infertile soils, ranching is intrinsically uneconomic in Amazonia. Without governmental support ranching would only be profitable when cattle prices were high and land values were increasing rapidly. If there is no increase in land values and taking the input of all resources into account, a positive rate of return simply is not feasible.

From a rancher's perspective, overgrazing is in all situations more profitable than proper (sustainable) land use. If the analysis is, however, extended to the use of all resources, overgrazing is far less preferable. Only if land values do not improve and cattle prices remain low will the advantages of overgrazing outweigh appropiate grazing intensity. In

Table a.1 Internal rates of return to a typical SUDAM-approved livestock project under two scenarios

Scenario[1]	Increase in land value (%)		
	0	15	30
High cattle prices			
Appropriate grazing intensity			
corporation resources (2)	16	18	24
all resources (3)	-1	2	9
Overgrazing			
corporation resources	23	24	27
all resources	-2	0	4
Low cattle prices			
Appropriate grazing intensity			
corporation resources	-3	6	17
all resources	-14	-6	5
Overgrazing			
corporation	16	18	23
all resources	10	-7	-1

Source: Mahar, 1989.

Notes:

1) Low input prices are assumed in both scenarios.
2) Ignores capital expenditures financed through fiscal incentives and official credit.
3) Fiscal incentives and official credit treated as if they were corporation's own capital.

this worst-case scenario the internal rate of overgrazing will be less negative.

The reason for Hecht et al. to analyse the consequences of increasing land values stems from empirical observations in the 70s and 80s, when land prices rose faster than overall inflation. It is unlikely that this situation will persist, however, because the underlying causes, the extremely high inflation rates and the irrational fiscal policies, are bound to change.

In general it can be argued that cattle ranching is inherently uneconomic and serves as a large burden for the Brazilian tax payers. It seems that cattle ranching was a 'bet on the wrong horse' which brought Amazonia environmental degradation instead of economic prosperity. Most of the current agricultural activities in Amazonia can only be realized by governmental support, which up till now has turned out to be socially unworthy. This is in contrast with the high profits which individual farmers and ranchers can realize by land expansion.

The Brazilian government, however, seems to have been rather sensitive to the demands of special interest groups, especially if these represent a powerful elite. To abandon the present fiscal system abruptly would be a form of political suicide. The situation is changing, however, because since 1979, livestock projects in the rainforest (floresta densa) are officially prohibited, although the enforcement of this measure has proved to be difficult (Mahar, 1989).

Market imperfections and the expansion of land-clearing activities

The economics of deforestation in the Amazonian area are intriguing indeed. Given the prevailing conditions in Legal Amazonia most players in the past forty years have been trying to act rationally, i.e. selfishly. Under some strict conditions such as perfect competition, the presence of normal goods, well defined property rights and no externalities, selfish behaviour should lead to a Pareto optimal situation. When the current 'chaos-and-disorder' economy in Legal Amazonia is considered, serious doubts arise about its optimality. Somehow these conditions in Amazonia must be such that convergence to Pareto optimality is prohibited.

One of the most important factors for an efficient allocation of resources in a well-functioning economy is a good property rights structure. An efficient property rights structure has four main characteristics (Tietenberg, 1988):

1) **Universality:** all resources are privately owned and all entitlements completely specified.
2) **Exclusivity:** all benefits and costs accrued as a result of owning and using the resources should accrue to the owner, and only to the owner, either directly or indirectly by sale to others.
3) **Transferability:** all property rights should be transferable from one owner to another in a voluntary exchange.
4) **Enforceability:** property rights should be secure from involuntary seizure or encroachment by others.

None of these four prerequisites is held in Legal Amazonia. The region faces an enormous open frontier, which expands rapidly by spontaneous land clearing. Virgin forests in, for example, Rondônia are treated as common goods, in which land clearing becomes a real life version of catch-as-catch-can. The possession rights can be acquired by official application or by land improvement, leading to conflicts whenever two different persons claim to be the legimate owner of the same parcel. The fact that the universality preposition is not held makes the scarcity rent of a forest disappear. The threat of seizure decreases the storage value of the trees on uncleared tracks of forest.

Marketable trees are logged and sold at prices which, from an intertemporal efficiency perspective are too low. The remaining unmarketable trees are burned. It is expected, however, that many currently unmarketable trees will be marketable in the future. The supply of tropical hardwood from Asia will stagnate, because those forests are facing severe overexploitation at the moment. In an efficient situation forests would be used up to the point where marginal benefits equal marginal costs and the scarcity rent will be maximized. In Figure a.4, this situation can be visualized: at the price, P*, an area of Q* will be cleared and the scarcity rent therefore becomes the rectangle ABPmP*.

However, since forests are treated as common goods, they are used up to the point where average benefits equal average costs, and are therefore cleared at Qm (see Figure a.4). Nobody has the unique ownership over the virgin forests, which motivates everybody to try to acquire as many of these forests as possible. This rational behaviour eliminates the scarcity rent and leads to excessive forest destruction (Qm instead of Q*). Furthermore, Amazonian land use systems do not only fail to fulfil the exclusivity preposition with respect to common goods, but also with respect to the attendance of 'small-scale' externalities, which occur in the process of obtaining fields and pastures. When a potential farmer clears a track of forest, he will not consider the costs of this behaviour: the effects in terms of erosion and climatic change, due to forest clearing, on adjacent and distant agricultural lands. When such negative externalities occur marginal private (MCp) and marginal social (MCs) costs start to diverge. This will lead to a discrepancy between the actual market equilibrium and optimality.

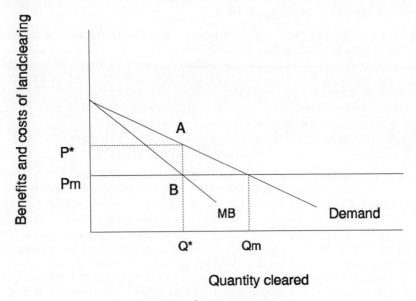

Figure a.4 Common goods

In Figure a.5, one sees that too much land is cleared (Qm instead of Q*) and also that the price of land, or at least the cost of land use, is too low (Pm instead of P*). This means that too many negative externalities are 'produced' and that there are too few incentives for the land users to search for land-saving techniques.

The prevailing situation is not efficient and should be dealt with by the authorities. A theoretical solution would be the adoption of taxing policies to discourage large scale agriculture. If, for instance, land taxes are set at (P* – Pm)/Pm an efficient solution comes about. When no other market imperfections exist, the implementation of such a so-called 'Pigouvian' tax would indeed lead to Pareto optimality.

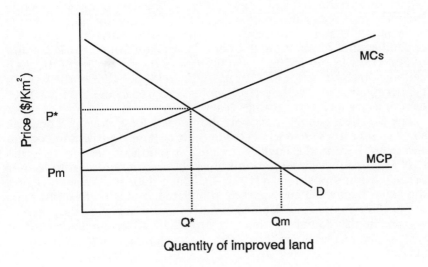

Figure a.5 Negative externalities

PUBLIC POLICY IN AMAZONIA FROM 1912 UNTIL 1992

An historical background

In order to understand the reasons why the settlers developed Amazonia in the way they did, that is by large-scale farming, one has to take the historical developments of agriculture in Brazil into consideration. The development of sugar in the 16th century and the development of coffee in the 19th century largely took place by land extensification, turning almost all the fertile areas in Brazil into plantations. This past extensification of plantations had two important implications for the recent development of Amazonia. First, because of those extensifications all the fertile parts in the moderate climate zones were already in productive use before the beginning of the Second World War, leaving the land-seeking masses no option but to reclaim the tropical forests.

Second, the historical development of Brazilian agriculture was a precedent for further development. That is, large-scale farming is embedded in the Brazilian culture: large-scale cattle ranching was supposed to develop Amazonia in the same way as the sugar plantations developed the Northeast in the course of the 16th century and the coffee plantations the Southeast in the 19th century.

Another important historical process that influenced the economic development of Amazonia was the rise in the demand for rubber. At the end of the nineteenth century a spectacular export boom of rubber took place in the Amazon region to meet the entire international demand. This was because Amazonia had at that time a natural monopoly in the extraction of rubber juice.

The increasing demand for rubber in the emerging automobile industry pushed rubber prices upward, inducing poor people especially from the Northeast to move to the Amazon region and expand rubber production. Rubber exports rose from a yearly amount of 6000 tons in the 1870s to 35,000 tons in the first years of the 20th century. In the latter period Brazil supplied 90 per cent of the world's rubber production, and by 1910 rubber exports accounted for 40 per cent of Brazil's export earnings (Baer, 1983).

The Brazilian rubber monopoly did not last long, however, because some rubber trees were smuggled into the botanical gardens of London and from there to the English colonies in Southeast Asia. In 1899 the first Asian rubber appeared on the world market, and Asia took over Brazil's leading position by 1912–13. Since then the Amazon region remained backward and sparsely populated, and therefore of minor importance to the Brazilian economy. Instead of ignoring the Amazon region, in the course of time the Brazilian government developed various plans to speed up the economic development of the Amazonian jungle.

THE RUBBER SUPPORT PLANS

First of all, public policy in Amazonia was introduced to revive the spectacular rubber boom, that lasted from 1870 until 1912. In 1912, a rubber support plan was initiated to try to maintain Brazil's market share. Premiums were given for planting rubber trees, for building rubber-processing plants, and for hospitals, railroads, port facilities, and housing. Import taxes on inputs such as fuel and merchandise for the production of rubber, fish and livestock were abolished. Export taxes on rubber were also reduced by the government. However, lack of financial means, as well as the impossibility of meeting the competition from Asia, doomed the plan to failure, and it ended in 1913.

The second rubber plan was the 'Batalha da Borracha', the battle for rubber, during the period 1942–1947. This 'battle' could not turn the tide either, although it lasted longer and had some (short-term) success, because of the negative impact the Second World War had on the rubber exports from occupied Asia to the Allied Forces. The end of the Second World War, however, meant the final defeat of Brazil's rubber-plan. By then, it had also become clear that Amazonian development could not sufficiently be stimulated by focusing on rubber alone. This approach was therefore succeeded by a much broader concept of regional development: greater efforts were made to initiate investments not only in agriculture, but also in industry, in infrastructure, and in health care, regardless of what the ruling regime in Brazil was.

1946–1964

After the loss of the 'battle for rubber' the Brazilian government founded a coordination agency, called the SPVEA (Superintendência do Plano de Valorizaço Econômica da Amazônia). This agency was situated in Belém and was supposed to supervise the regional development plan, which at that time was still to be developed.

The first achievement of SPVEA was a clear specification of the region for which SPVEA's own founding was justified. Not only the area of 'classical Amazonia' (Amapá, Acre, Roraima, Rondônia, Pará, and Amazonas) was covered by SPVEA, but also parts of Maranhão, Mato Grosso, and Goiás. The new classification was based merely on economic interrelationships, instead of on geographical and administrative borderlines, as was the old classification.

The first five-year plan saw the light in 1955, giving attention to various objectives including the promotion of agriculture, industry and mining. The plan also developed programmes dealing with flood control, transportation, communications, energy, social welfare, banking and credit, and general research.

The results of the rather ambitious first five-year plan were modest, due to the lack of scientific foundations of the plan, combined with financial and administrative problems within the SPVEA. Although the SPVEA had, in theory, the legal right on 3 per cent of federal tax revenues to make sure that it would have enough funds for the implementation of the necessary investments in the years to come, in practise it did not work, because the financial promises could not be kept by the government. A positive side effect of the plan's failure was its consequential limited impact on the exploitation of the tropical forest.

The major accomplishment of the development programme of 1955–1960 was the construction of the Belém–Brasilia highway, opening the states of Maranhão and Goiás. The population in those areas rose from 100,000 persons in 1960 to two million persons in 1970, 'improving' the forests on both sides of the road into agricultural land.

Operation Amazonia

In 1964 the military took over in Brazil. The old development plans were abolished, giving opportunities to new ones. In line with military tradition, the new plan was called 'Oparação Amazonia', suggesting similarities between occupying Amazonia and an arbitrary hostile country. One of the main reasons for the military to continue the active stimulation of the development of legal Amazonia was to ensure Brazil's sovereignty in that region. Venezuela and Peru made great efforts to develop their parts of the Amazon basin, making the sparsely populated confines of Brazil quite vulnerable.

The first reason for the military to develop Amazonia was thus based on geopolitical motives and hoped to promote the building of self-supporting settlements along the immense Brazilian borderlines. The second reason for continuing the aid to Amazonia was the firm belief that sooner or later this would bring benefits to Brazil as a whole. The new regional policy was based on import substitution. With the aid of capital from South Brazil and foreign capital Amazonia was able to diversify its economic structure.

Public policy was aimed at improving the infrastructure and the fiscal climate, not only for Amazonia as a whole, but especially for certain development poles in the inlands of Amazonia. A good example is Manaus, the capital city of Amazonas, which was turned into a free trade zone, leading to the greatest revival of Manaus' economic growth since the collapse of the rubber monopoly.

A new institution was founded to replace SPVEA, which had not been functioning well according to the new government. In the period 1964–1966, less than 5 per cent of SPVEA's funds were invested in western Amazonia, which was too little to ensure national sovereignty. Most investments were made in the subregion of Belém, due to its superior infrastructure and larger market. The SUDAM (Superintendência do Desenvolvimento da Amazônia), as this institution was called, was linked to the Ministry of the Interior and shared a lot of similarities with SUDENE, the development agency for the depressed Northeast. And just like SUDENE, SUDAM funds were administered by a regional development bank. For Amazonia this was BASA (Banco da Amazônia).

The first five-year plan of SUDAM contained two priorities:

1) The development poles had to be connected with each other. Roads were therefore planned to be built between Porto Velho and Manaus, between Manaus and Boa Vista, and between Cuiabá and Porto Velho.

2) Modern agriculture and industry had to be stimulated. The new government believed that import substitution was a panacea for unprecedented growth.

The achievements of the first five-year plan were disappointing, however: not only did the number of actual investments stay behind the expected number, because SUDAM did not control all the projected investments (only 12 per cent of planned investments were to be financed with SUDAM funds, the other 88 per cent with private money or with money from other governmental departments), but their productivity also fell short of expectations. Furthermore, it also turned out to be politically difficult to stimulate West Amazonia at the expense of Belém, which still attracted a lot of private capital. An effective regional development for East as well as West Amazonia could only be realized by expanding the role of the government. At least, that was the government's own opinion.

This led to the development of the national integration programme (PIN). Before this programme is discussed, however, first some attention will be given to the fiscal system of SUDAM and the appearance of livestock in Amazonia, because of their important role in the deforestation process in that region.

The fiscal system of SUDAM and the appearance of livestock in Amazonia

In 1964 Brazil faced a coup d'état, leading to radical changes in regional policy. One of the most important changes for Legal Amazonia was the establishment of SUDAM to coordinate and to finance the import substitution industries. SUDAM supervised a fiscal programme, which hopefully would stimulate industries to choose their location within the Amazonian boundaries.

In order to acquire money to finance the import substitution industries, the Brazilian corporations could take up up to a 50 per cent credit against their federal income tax liabilities if they invested the resulting savings in SUDAM-approved projects in Legal Amazonia. These investments could be made in new projects or in modernization and expansion of existing projects. In the beginning only industrial projects had access to those investment tax credits. Since 1966 agricultural, livestock and service sectors could also apply. Depending on the priority given by SUDAM, investment projects could be financed up to 75 per cent with tax-credit funds. Factors such as employment creation, export potential, the use of innovative technologies and the choice of location determined the priority and thus actual amount of credit offered to investors.

The SUDAM tax credit system did not oblige suppliers of tax credit to invest in their own enterprises, e.g. to create their own demand, but also offered them the opportunity to deposit the money at BASA. The BASA thus functioned as an intermediary between demand and supply of SUDAM funds. Consequently the possibility emerged that demand would not equal supply. The demand for tax credits depends on the profit expectations of the investors, which in its turn partly depends on tax credit policies in the future. The supply of tax credit depends on the present profit, which is influenced by the overall economic situation. Up till 1971 supply exceeded demand; thereafter the opposite was the case. The increasing demand for tax credits led to brokerage fees of over 30 per cent (Mahar, 1979).

In the period 1964–1985 more than 950 projects had been approved: over 300 in industry, over 600 in livestock, and about 30 in services. Most of these projects were approved in the late 60s/early 70s, peaking in 1970. Until 1970, most emphasis was given to industrial projects in the import substitution sectors. After 1970 the subsidization of livestock projects became the most important task of SUDAM. The first livestock projects were approved in 1967. Since 1971, more than 160 modification projects in the livestock sector had to be approved to try to convert loss-making ranches into profitable ones again.

By 1985 631 livestock projects were approved, converting 8.4 million hectares of forest into pastures. This was responsible for about 10 per cent of total deforestation in Legal Amazonia (Mahar, 1989). Most of the SUDAM-financed ranches, some 75 per cent, were located along the highways in Pará and Mato Grosso. Over the years, livestock projects consumed 40 per cent of all tax credit funds. SUDAM spent 700 million dollars on livestock, more than one million dollars per project.

Although tax credits by SUDAM gave rise to the founding of many ranches in Pará and Mato Grosso, this was not the case in the other parts of Legal Amazonia. Moreover, probably some 90 per cent of all cattle ranches were established without SUDAM money (Mahar, 1989). However, this does not mean that these ranches were not somehow supported by the Brazilian government, and still are. Rural credit is offered at a real negative interest rate, functioning as an implicit subsidy. Furthermore, landholders cannot only depreciate their entire fixed capital in the first year, but are offered the opportunity to depreciate it several times, up to a multiplication factor of six (Binswanger, 1991).

The emergence of livestock in Amazonia is sometimes blamed on the worldwide increase in the demand for beef. This so called 'hamburger connection' between South America and especially the United States of America is indeed a controversial issue (Shane, 1986). Many Central American countries are net exporters of beef, making the connection plausible. The Amazon region itself, however, remains a net meat importer (Hecht, 1985; Sawyer, 1990).

The National Integration Programme

In the early 70s, regional policies changed once again. The support for import-substitution industries was abolished, because of the government's inability – due to a lack of financial means – to make long-term commitments to those industries, and try to let them obtain a reasonable market share.

By that time, the government believed that Amazonia had to be integrated into Brazil's economy in such a way that its comparative advantage could be exploited. This meant that the abundancy of still-to-be-improved land in Amazonia had to be combined with the severe unemployment rates in the other parts of Brazil. It was expected that massive migration from the Northeast and Southeast to Amazonia would bring benefits to all regions.

The national development programme PIN (Programa de Integração Nacional) was assisted with PROTERRA (Programa de Redistribuição de Terras), a land distribution programme. According to PIN, two more roads had to be built, one connecting the East with the West (the Trans Amazon Highway), and one connecting the North with the South (the Santarem-Cuiabá Highway).

All the land along those roads had to be redistributed by PROTERRA. The objectives of PROTERRA were:

- promoting small-scale agriculture by offering rural credit;
- supporting investments in agro-industry;
- subsidizing modern agricultural techniques; and
- supporting output prices.

The PROTERRA programme was supervised by the Ministry of Agriculture and by the INCRA (Instituto Nacional de Colonização e Reforma Agrária).

The direct results of the official immigration plans were disappointing. According to the plan, half a million peasants were enabled to join the official settlements, while only one-tenth actually appeared. Reasons for this meagre result were the long-term impossibility of establishing annual cropping due to the poor soils, the lack of storage facilities, and the lack of technical assistance. Although there were few official settlements, spontaneous immigration to Amazonia was considerable. Poor people invaded large parts of accessible land acquiring possession rights for the land by improving it. In practice, this led to shifting cultivation practices, causing massive deforestation.

The oil crisis of 1973 made national integration on the basis of highways, trucks and personal cars a particularly uncomfortable policy option. Massive oil imports worsened the trade balance, increasing Brazil's external debt. At the same time, it became clear that PIN could not solve the demographic and socio-economic problems of the Northeast, which made the programme a mixed blessing.

The aim of the national integration programme was to treat Amazonia as a 'resource frontier,' which had to be exploited for the benefit of the entire country. Therefore, it had to be investigated what the quantity and quality of the available resources in Amazonia really were.

Although some earlier surveys were already carried out under the supervision of SPVEA, the new techniques were much more sophisticated and therefore more accurate. The entire Amazon region was aerial mapped by RADAM (Radar na Amazônia). Through photo-interpretation and complementary field studies, RADAM was thus entrusted with elaborating the first systematic inventory of minerals, soils and vegetation ever attempted for the entire Brazilian Amazon.

POLAMAZONIA and the greater Carajas Programme

The discoveries of RADAM were impressive. In the state of Pará, large iron ore reserves were found with promising perspectives for exportation. The Brazilian government once more changed her Amazonian development programme. National integration was substituted for spatial development. The new programme was called POLAMAZONIA (Programa de Pólos Agropecuários e Agrominerais da Amazônia). The programme of Agricultural, Livestock and Mineral Poles in Amazonia emphasized the development of large scale, export-oriented projects in the livestock, forestry, and mining sectors in fifteen 'growth poles' scattered throughout Amazonia. Exports of minerals, timber, and agricultural products could potentially earn a lot of foreign exchange, which was badly needed to meet Brazil's interest payment obligations.

The most important part of POLAMAZONIA was the exploitation of the discovered iron ore reserves in Pará, or more precisely, in the subregion of Carajás. The Carajás mine contains the largest reserves of high-grade (66 per cent) iron ore in the world: an estimated 18 billion tons. At an extraction rate of 35 million tons per annum commencing in the late 1980s, iron ore reserves are predicted to last over 500 years. Other significant deposits include manganese, copper, bauxite, nickel and stannum (Anderson, 1990b).

In order to exploit those reserves rationally, the Brazilian government established the Greater Carajás Programme (PGC) in 1980. The programme area includes 895,000 square kilometres – more than 10 per cent of Brazil's total land area. Within this area approved firms are offered fiscal incentives in the form of tax holidays and exemptions from import and export tariffs, credit guarantees, and subsidized electricity from the Tucurui hydroelectric facility (Binswanger,1991; Anderson,1990b).

The cornerstone of PGC was the Carajás Iron Ore Project, which consisted of three complementary investment projects: the realization of the iron mine itself, the improvement of the port of São Luis, and the building of a 780 kilometres long railroad through Pará and Maranhão connecting the mine with the harbour. The iron ore project was supervised by the Companhia Vale do Rio Doce (CVRD), a state-owned mining enterprise.

The ecological consequences of this part of PGC are negligible. The same cannot be said, however, about the other parts of the plan. Private investments in mining companies, charcoal industry, pig-iron industry, and agriculture will have devastating effects on the tropical rainforests in Pará and Maranhão.

First, the massive migration flow to the new 'El Dorado', due to the job opportunities in the construction sector and the discovery of gold, caused a population increase which has led to violent conflicts over land rights and resulted in an intensification of the environmental degradation in that region. Second, large parts of the rainforest will be logged for the essential supply of timber for charcoal production. This charcoal is needed to transform iron ore into pig-iron, which is easier to transport and can be sold at higher prices.

The profitability of the pig-iron industry depends mainly on energy and pig-iron prices. As long as energy prices are lower than $70 per ton of charcoal equivalent, and pig-iron prices presumed $100–$110, pig-iron plants will be profitable (Mahar, 1989). Since charcoal from virgin forests is sold for $27 per ton, the plants will be very profitable indeed. However, not all the costs of charcoal production are reflected in its price: only the cutting and transportation costs of logging activities are borne by the charcoal producers and are therefore passed on to the pig-iron industries. However, the other environmental costs are externalities which are not included in its price. There is, therefore, a divergence between private and social costs, making the charcoal price of $27 per ton a bad indicator for the future perspectives of the iron-ore industry. Some attention was given to the market solutions in the case of negative externalities in earlier chapters.

It is estimated that each year 1.2 million tons of charcoal are needed to provide all planned pig-iron industries with this input. This implies that, on a yearly basis, 90,000–200,000 hectares of virgin forest have to be logged (Mahar, 1989). Although the possibility exists to extract timber from plantations, this will take a long time to implement. Only after a period of 20–30 years can the supply of timber from virgin forests, in theory, be replaced by the supply of sustainably produced timber, because sustainable rotation schedules can only be reached when rotation periods of at least 25–30 years are introduced (Anderson, 1990b).

As already mentioned, it is highly questionable, however, whether the supply of timber from sustainable silviculture can replace the demand for timber from virgin forests. According to Fearnside (in Mahar, 1989), an area of 2.6 million hectares of fast growing eucalyptus trees is needed to assure a sustainable timber supply for the charcoal industry. There is no proof that plantations of such a large scale can be sustainable in Amazonia at all, because of its bad soil fertility and its harsh climatic conditions. Even if the implementation of such immense plantations is ecologically possible, which is highly uncertain at best, then practical aspects will still make this a rather Utopian solution. First, small profit margins and unstable economic perspectives make long-term planning and risk calculation impossible, or at least awkward. Unfortunately, because of the long rotation schedules, sustainable silviculture usually does require stable future expectations. As long as the circumstances for logging activities in the tropical forests do not change, that is as long as the loggers do not have to pay for the environmental damage, timber from silviculture cannot compete with timber from virgin forests.

Second, the eucalyptus plantations are likely to be occupied by poor squatters, who cannot afford to buy land at other locations due to the ever increasing prices of land. The charcoal producers, who own those occupied plantations, have two options: they could try to protect their plantations or they could leave them and ask the Brazilian authorities for new ones, starting the cycle again. Until now it was easier and more profitable for the plantation owners to acquire new tracks of forest, because in most cases the standing volume of those new tracks of forest was far greater than the standing volume of the old occupied plantation.

Some argued that if no reforestation takes place, the charcoal industry will destroy the entire tropical forest of Brazil over 70 years (Anderson, 1990b). The overall impact of the Greater Carajás Programme on the environment therefore seems to be negative.

POLONOROESTE

As was mentioned before, highway building was seen as of utmost importance for the development process of Amazonia, because those roads made the jungle cities far more attainable and gave access to millions of hectares of 'unused' land. The former inaccessible land loses its natural protection against human occupation, leading to the conversion of large tracks of virgin forests into pastures and fields. In this sense, the construction of the Cuiabá-Porto Velho highway (BR–364), which was part of Operation Amazonia, was extremely successful.

In 1960, before the road was built, only 70,000 people lived in the 243,000 square kilometre state (then a federal territory) of Rondônia. Most of those people were rubber tappers, which explains why by then the rainforests, which covered 80 per cent of the state, were still intact. In 1968, after the completion of the BR-364, migration to Rondônia increased drastically from perhaps 3000 persons per year in the 60s to a tenfold of that number in the 70s. Reasons for the willingness of people to migrate into Rondônia can be subdivided into various push and pull factors.

The most important push factor for the rural people in the South to leave that region was the decline in job opportunities, caused by the mechanization in soybean and wheat production. Other push factors were the bad harvests in the coffee sector, due to killing frosts, and the fragmentation of landholdings in the Central-South.

The appreciable pull factors for immigrants to go to Rondônia were first, the good fertility of some of the soils along the highway, which made the opportunities for agriculture very promising. Because this information was published by the government and confirmed by the early settlers through letter-writing to their families, this initiated a chain-migration. However, the conditions for agriculture often turned out much less roseate than expected, because the fertility of large parts of Rondônia's soils had been exaggerated. Second, the Brazilian government offered each new settler a parcel of 100 hectares at a nominal price, together with basic services and infrastructure.

The increase of inhabitants in Rondônia, a net inflow of 30,000 persons per year, had devastating consequences for the area of tropical forest turned into agricultural land, for unfortunately, the new farmers did not try to implement permanent agriculture on relatively small parcels of land, but alternatively, used shifting cultivation methods to ensure minimal soil degradation and therefore good harvests.

It is argued that forest fallow is the most efficient sustainable land-use system as long as population density is very low (Boserup, 1981), making the disappearance of tropical virgin forests along the roads inevitable, especially when the rates of return on those maintaining forests are undervalued. However, in the period 1975–1980 the network of feeder roads increased fivefold, while in the same period the area deforested increased sixfold (Mahar, 1989).

The combination of socio-economic problems in Rondônia and the bad condition of the BR–364 during the rainy season made the government decide that a special integrated programme was needed to cope with both problems at the same time. The programme, which was expected to bring a solution, was the Northwest Brazil Integrated Development Programme, or POLONOROESTE, which covered the entire state of Rondônia and parts of western Mato Grosso. The programme's main objective, apart from paving the road, was to reduce forest clearance on land that was expected to have no long-term productive potential. Furthermore, the programme tried to stimulate sustainable agriculture based on tree crops. It is obvious that the plan's two objectives conflict – preserve the forest by making it easier to access! – and that therefore at least one of them may be sacrificed.

The improvement of the BR–364 in 1984 attracted a lot of new immigrants: 160,000 persons per year in 1984–1986 compared with 65,000 persons per year in the period 1980–1983. Due to the massive increase in population, the pressure on the tropical forest accelerated, turning POLONOROESTE into a failure. In addition to the population problem, several institutional factors also stimulated, or at least did not try to discourage, deforestation. First, the Brazilian Institute of Forestry Development (IBDF) has not been able to enforce the 50 per cent rule to try to reduce the pressures on the forest. Although farmers are not allowed to clear more than half of their holdings, most of them did. In fact, precisely due to this rule one runs the risk of increased dispersion of both agricultural lands and tropical forests, leading to unsustainable forests, because the plots will be too small to guarantee a sufficient biological diversity.

Second, the abundance of land in sparsely populated areas makes fallow systems more efficient, and therefore more rational, than annual cropping on permanent fields. In 1987 the population of Rondônia was estimated at 1.2 million, or on average five persons per square kilometre. According to Boserup (1981) this will make bush fallow the most rational choice for farmers in low technology countries. If the government, however, still wants to encourage annual cropping and tree cropping on permanent fields, which are both more labour- and capital intensive than bush fallow, it must give the farmers financial incentives in the form of subsidies and cheap credit to make the use of chemical fertilizers a sound alternative for them. The INCRA could not carry this financial burden, because the Brazilian government was obliged to adopt austerity measures to cope with their huge financial problems in 1983.

Third, the INCRA accepts deforestation as a sort of 'land-improvement', which not only by ancient but also modern Brazilian law, gives the 'land-improver' the possession right, the *'direito de posse'*, for three times its cleared area up to a maximum of 270 hectares. This law seduces farmers to clear more forest than they actually need or have the means to cultivate properly, since they can make large short-term profits by selling some of the legally acquired plots. In conjunction with the ever-increasing land prices this turns agricultural landownings into speculation goods, and makes today's productive use irrelevant.

Fourth, and last, the rural land tax system stimulates deforestation, because it gives farmers a reduction of up to 90 per cent on their tax payments if the land is in productive use. Unfortunately, deforested land is always seen as productive. The impact of the rural land tax system on deforestation is small, however, because the tax burden is low: the tax percentage is only as small as 3.5 per cent of the market value of the land, is declared by the farmers themselves, and can therefore be manipulated to minimize actual tax payments.

CONCLUSIONS

In this case a description is given of the public policies in Amazonia which tried to stimulate the economic growth of that region. Most of these policies did bring some short-term benefits to the Amazonian people, but only at the cost of considerable expense for the Brazilian society as a whole. In fact, the public policies could only have been welfare-improving for the entire Brazilian economy, if the government had had an information lead over the producers and consumers concerning the economic potential of the Amazon region.

It does appear that up till now, this most likely has not been the case for Brazil. The reallocation of labour and capital from the South to the North brought neither Amazonia nor Brazil any long-term prosperity. The government seems to have misjudged the agricultural potential of the jungle, at least under current techniques and lack of fertilizer input.

Therefore, at the heart of the current problem lies the historical governmental wish to integrate Amazonia into the national economy, which unfortunately was done without enough inquiries into the ecological feasibilities of land-clearing activities in the jungle.

Environmental awareness did not show up in official plans until in the course of the 70s, when much damage had already been done: in the 60s and 70s the Brazilian government focused on economic growth, and on trying to cope with poverty. The simple belief was that the exploitation of the enormous resource frontier would bring benefits to the rapidly increasing population.

Several highways were built to open up Amazonia, while at the same time agriculture in the 'empty' territories was stimulated by the provision of tax credits and subsidies. The direct causes of deforestation were therefore road building, cattle ranching and shifting cultivation, which were all, directly or indirectly, stimulated by the government. Although the subsidy system succeeded in converting tropical forests into pastures, the pastures' productivity has been rather low, among other reasons because of Amazonia's poor soil fertility. There are clear signs that the Brazilian government is about to diminish her financial support to these uneconomic and unsustainable land-clearing activities.

One of the underlying motives for some Brazilians to migrate to Amazonia is their wish to escape from their state of extreme poverty. The poor landless people have virtually no option but to occupy the marginal lands in the tropics, or to move into the overpopulated cities in the South. However, both alternatives offer little or no opportunities for them to gain any serious increase in wealth, which makes them often desolate and indifferent to long-term considerations.

APPENDIX 2

THE INDONESIAN CASE: LOGGING AND DEFORESTATION

FORESTRY POLICY IN INDONESIA: AN OVERVIEW

The above global analysis demands a further specification at the country level. Here we will try to narrow down the problem to the Indonesian case.

Indonesia contains about 8 per cent of the world tropical forests, and some 13 per cent of the world's commercial reserves. Moreover, at present some 20 per cent of all logging activities in the world take place in the Indonesian archipelago. In contrast with any developing countries Indonesia has favourable natural endowment. The country has a wealth of energy resources, in fact oil and gas provide its main export revenue. But essentially Indonesia is an agricultural nation, in this sector 60 per cent of the population is employed and it generates 25 per cent of GDP (Blom, 1991).

Indonesia became independent from Dutch colonization in 1949, after which followed a decade of political and economic instability. Only after 1965 did the Indonesian government achieve stability under Soharto power, and a policy strongly aimed towards development started. Forest resources, together with oil and gas production, were considered the sources for internal economic development. The new government succeeded in attracting a large volume of foreign and local capital in support of forestry development. This caused the exports of forest products, in the beginning mainly in the form of logs, to become one of the most important sources of foreign exchange of the country. A large number of concessions for logging activities were granted and profits were assured by tax exemptions. In a second phase government policies turned towards industrial development and from the late 1970s on several policy measures have been taken to also promote the processing of the tropical hardwood in Indonesia. In 1978 the exportation tax on logs was doubled and most sawnwood and plywood taxes were exempted. In addition from 1980 log exports were progressively reduced and in 1985 they were totally banned. As a result Indonesia succeeded in successfully capturing between 70 and 80 per cent of the world's (hardwood) plywood market. So, presently the wood industry makes an even stronger positive contribution to the whole economy than before.

This is not to say, however, that this economic sector is totally free from criticism. First, it has been pointed at the waste due to the inefficiency – by international standards – of domestic timber processing. In fact the export restrictions subjected tropical timber trade to many distortions. Generally the justification cited for these policies is that exporters are compensated by making the price of raw logs higher for processors in importing countries, reducing the cost disadvantage for internal processors. This strategy has the dual purpose of reducing harvesting pressure on forest, by increasing value added per log extracted; and at the same time to create export revenues and employment.

But such a tax structure created a high rate of protection for internal wood-based industries and, in addition to price subsidies, as a result prices of wood are at a lower level than their real value. This underpricing of natural resources reduced incentives to make efficient use of forest resources. For plywood manufacture protection has been calculated to be in the order of 222 per cent (Barbier et al., 1991); and saw mills in Indonesia recover about 43 per cent of logs compared to 55 per cent in comparable developing countries; it means that in efficiency output could be increased by 28 per cent by the same log volume (World Bank, 1990a). In addition to depressing prices, the log export bans have led to the substitution of wood for other factor inputs. Thus the protection to Indonesian mills not only increased the log demand, but the operational inefficiencies ensured that more logs were harvested than if, instead of a forced industrialization, a policy to develop domestic processing capabilities had been implemented.

Another problem related to the affected structure of wood prices is that valuable species are not reserved for high-quality uses since they are not priced high enough; so for example valuable dipterocarps (meranti) are used for plywood cores (World Bank, 1990a).

Second, logging practices have been blamed for their adverse environmental impact. In Indonesia since 1969 logging activities are regulated by a system for the selection of cuttings and for silviculture replanting, in order to utilize forest resources on the basis of sustainability. The legal framework, renewed in 1988, is the so-called 'Indonesian Selective Cutting and Replanting' system (TPTI). The main practical guidance can be summarized in the definition of 35-years cutting cycle and in the obligation of concession holders to implement activities for the regeneration of forest.

This system had not the success it expected and instead had negative effects on forest resources, for instance due to the fact that the required rotation periods are not respected, or that unnecessarily large amounts of less marketable varieties are damaged through the loggers. Recent surveys show that up to 40 per cent of standing stocks are damaged in logging operation and where the rotation period is not respected the damage is even higher (Atlanta/INPROMA, 1987).

The 35-year harvest cycle is not compatible with the length of concessions periods (20 years) and in fact concessionaires are led to have a short-term perspective. Often they relog before the cycle is completed, again because legal sanctions are not very strong. In addition the most valuable species (dipterocarps) reproduce slower and under specific conditions. In this way the selective cutting system tends to protect more robust and less valuable species, altering the quality and the richness of the forest, which is especially valuable in Indonesia.

Logging activity also indirectly contributes to the transformation of forest; any kind of activity has the effect of 'opening up' unexploited forests. This is due to the fact that along with logging, other forms of economic activities must begin. For example developing new roads means new lands opened to smallholders and providing work opportunities through logging means attracting new families into the forest.

The Ministry of Forestry of Indonesia divided the total forested area into five classes:

1) Protection forests;
2) Parks and Reserves;
3) Limited Production forests;
4) Non-convertible forests; and
5) Conversion forests.

Parks and reserves cover 18.7 million hectares, which means that about 10 per cent of all Indonesian land area is set aside for conservation and protection. This a very large area in comparison to that so designated in most developed and developing countries. The main constraint in the reserve system is the absence of incentives for local people in the preservation of natural resources. In fact generally people have no awareness of environmental protection and, essentially, they gain no benefit from those protected areas.

The process of decline of the forest resources is further aggravated by deforestation, which is related to land use and its distribution. Indonesia has a population of 176 million and in 1970 the crude birth rate was of 44 births per 1000 population per year; about 625 per 1000 of the population is located in inner islands (Java, Madura, Bali and Lombok) that cover only 8 per cent of total surface area, so that for example in Java the population density is about 800 people per km^2 (World Bank, 1990a). The Indonesian government has an active programme for family planning, but it is difficult to reduce the growth rate because most poor households desire more than two children for income security. So in order to resolve the problem of population it is relevant to look at the income level.

Policy measures have been taken in order to reduce poverty and population pressure, promoting migration into the less densely populated regions, the outer islands. The outer islands contain about 98 per cent of all forested area in Indonesia. Between 1980 and 1986 more than 2 million people moved from the inner to outer islands and in order to support the transmigration programmes about 800,000 ha of land were cleared; of this about 300,000 ha were in brush and secondary forest and an equal amount was in logged-over primary forest, the remained was not forested. Although 40 per cent of Indonesians live in outer islands only 20 per cent of medium- and large-scale industries are located there. So demand for agricultural land is very strong, and creates pressure on forested lands.

In fact local smallholders and migrants have few incentives to preserve the forest. First, generally they face serious constraints in obtaining legal land titles because of the land tenure and shortage of cash; in addition they have not the means to use modern technologies and newly-cleared lands are more productive. As a result they are attracted to Forestry Department lands. In 1983 the Basic Forestry Law (BFL) was formalized and expanded; so that when this law was passed in 1967 about 26 ha of land were within Forestry Department boundaries, and in the reclassification the amount was 113 ha. Including the areas for potential conversion to other uses, about 75 per cent of the country falls within Forestry Department boundaries; this situation creates serious problems in acquiring land.

Second, the Basic Forestry Law recognizes the rights of local people to harvest forest products, except timber. Only concession holders have the right to harvest timber and they tend to discourage the exploitation of secondary forest products for clearing and poaching concerns. It means that smallholders have low interest in protecting the forest, so the main way local farmers can realize profits is by clearing Forest Department lands.

As a consequence of the declining of oil revenue in the 1980s, the rates of growth and employment of the Indonesian economy slowed, and agriculture had to absorb the surplus labour. During 1980–1985 the agriculture sector absorbed about 42 per cent of new entrants into the labour force. But agricultural production, especially in transmigration sites, is mainly based on low-input food production and in low-fertility soils. This situation, combined with land constraints, forces transmigrants to adopt shifting cultivation methods, thereby increasing damage to the forest.

Agriculture, in response to population growth, and forestry management, in response to ecological concerns, needs a better land allocation in order to ensure optimal land use from both the economical and environmental point of view. In this sense Indonesia lacks an efficient land classification. For instance, nearly 30 per cent of all land within Forestry Department boundaries in Sumatra is deforested but it is not available for development since it is classed as Forest Department land; some areas in Kalimantas are classed as production forest but, having a slope over 45 per cent, should be protected (World Bank, 1990a). Many agencies are involved in land use planning and land allocation, but because of the fragmentation of responsibilities and the lack of a lead agency for coordination a rational land management agreement has not been reached.

Other problems related to land uses come from development projects. In fact in theory provinces generally can select proposed developments on their territory but data tends to be centralized and not available to local decision-makers. In 1988, with the last Five-Years Development Plan REPELITA V, the Environmental Impact Assessment (EIA) was introduced in order to check all projects of intervetion on land. This system can be an efficient way to plan a rational development, according to the needs of population and of specific environments affected by the projects. But in actual fact most provinces are not ready, because of the lack of trained staff and of limited budgets.

Given Indonesia's prominent position in logging it stands to reason that the country can not escape from the main problems that are nowadays associated with the degradation of tropical forests: removal of the cover, reduced biological value of the remaining forests, the lack of control on land use by the smallholder, and finally deforestation processes. Presently deforestation in Indonesia makes up some 8 per cent of global deforestation: this was some 5 per cent during the first half of the 1980s (FAO, 1988).

Some parameters for Indonesia such as the policy-makers' time horizons have been brought in line with the projections made by FAO/GOI, 1990, and Lyons, 1990. Based on these studies, a time horizon of 35 years, corresponding with the institutional rotation period, seems most appropriate. Based on Ingram, Constantino and Mansyur, 1989, an additional change in the growth rate of removals to 4 per cent also was considered appropriate.

THE IDIOM PROJECTIONS FOR INDONESIA

The IDIOM projections on Indonesia's volume of logs are presented in Figures a.6–a.8; the absolute size is illustrated in Figure a.6; the relative size in a.7 and the destination in a.8. Figure a.6 first of all shows that the gap between the actual amount logged and what is required to satisfy domestic needs (Minlog) is expected to fall during the two next 15 year periods to a negligible amount. More specifically the present amount which is available – whether or not in a processed form – for exports is some 20 million m^3. In the year 2000 only 10 million m^3 will be available for exports, and in 2010 domestic needs will cover all logging activities in Indonesia.

By then this scarcity will also make itself felt in domestic markets as Figure a.6 indicates. All this will contribute to a serious decline in Indonesia's position as one of the main producers of tropical timber in the world. Figure a.7 suggests that the share in world production, now some 20 per cent, will decline within a decade to a level around 8 per cent. In the following years a gradual increase in Indonesia's position comes to the fore: in 2030 a level of 12 per cent of the world production is reached. The Indonesian role as a leading exporter of timber or processed wood products is also due to disappear throughout the next two decades. After 2010 its role in the world market is projected to be far less prominent than at present.

The question now comes to the fore, to what extent are the above projections in agreement with the views held in the country itself? The currently available volume of annual log harvests of tropical timber have been represented in Figure a.8. Their sources were the Dept. Kehutanan 1984 for 1960–84 and the Ministry of Finance for 1986–87

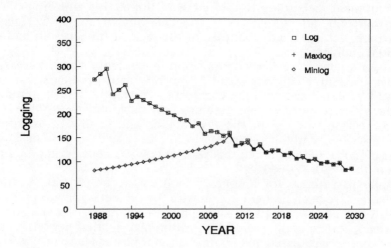

Figure a.6 Logged hardwood (10^5 m^3) – Indonesia

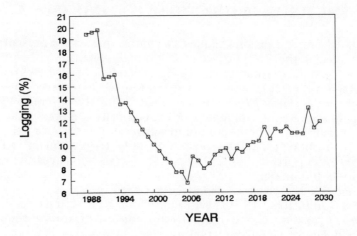

Figure a.7 Logged hardwood – share of Indonesia in world logging

(both represented in Repetto, 1988, p.44). To these the IDIOM figures were added for 1988–90[1].

Indeed the long-term perspective for Indonesia is that it cannot escape the fact that it will play a much less prominent role in the world timber market than at present. First, domestic supply will slow down

---

[1] The reader should be warned that different sources may well present data that vary up to 10 per cent from each other. Moreover there may be some differences in the definitions used in several sources.

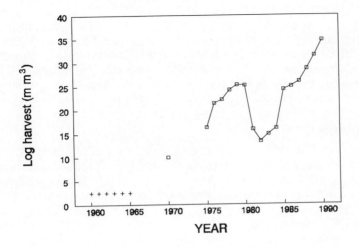

Figure a.8 Volume of annual log harvest (million m³) - Indonesia

according to our projections, but secondly, increasing domestic demand relative to the remaining commercial reserves will crowd out exports to such an extent that Indonesia's role as an exporter of timber may gradually be taken over, according to the IDIOM projections, by other producing countries such as Brazil and Zaire.

It should be mentioned here that by definition no definite statements can ever be made with respect to a period which is more than a decade ahead. In this case where the focus is on the future Indonesian market share, this is further aggravated because by then, through the disappearance of a strong market leader, the market will likely have become rather sensitive to the size of the remaining commercial reserves, and thus will be very unstable in terms of market shares. Our projections for the period from 2000 onwards indicate that most likely Zaire will become a new dominant supplier in the world market (with a market share above 40 per cent) along with Brazil (market share some 30 per cent).

Although it is hard to speculate about the impact of reduced Indonesian exports on the regional patterns of trade, some potential developments as suggested by the model results are worth mentioning. First, Indonesia's dominant position in the Japanese market, representing now some 15 per cent of world demand for timber (products), will be eroded. Currently some 70 per cent of Japanese imports of tropical timber originates from Indonesia; this share may be some 50 per cent by 2005 if Indonesia trades solely with Japan. Even then, however, this share will probably decline further.

Just as with the other main producing regions, Indonesia can not escape from both a fall in the commercial reserves and progressive deforestation. The projections indicate that, unless special policy measures are taken – note again that IDIOM already does recognize a

35-year conservation horizon for Indonesia – the commercial reserves of Indonesia in the world total will fall from 13 per cent now to some 10 per cent in the near future. At the same time deforestation is expected to rise excessively: from 0.7m ha per annum nowadays to more than twice as much in a few decades. If this process continues, in 40 years' time some 35 per cent of Indonesia's tropical forests will have disappeared, and the size of the farmland area (now some 30 per cent of the tropical forest area) will surpass that of the forest.

Some policy measures

To illustrate IDIOM's potential as a tool for policy analysis, some preliminary simulations were carried out. All policies are simulated in isolation and focus on the Indonesian case. They subsequently deal with: *a.* an import quotum imposed by a major tropical timber importing region (i.e. Europe); *b.* an import quota based on mutual agreement between all developed countries; *c.* a restriction on the size of Indonesia's exports; *d.* programmes to reduce population growth in Indonesia; *e.* a serious expansion of the Indonesian area for national reserves; *f.* measures by the Indonesian government to raise agricultural productivity.

a-b relate to international policy options, whereas *c-f* deal with measures which could be taken by the Indonesian government.

A European quota
In this case the assumption is made that through quota Europe will force down its total tropical timber (products) imports from 1988 on by half the 1987 level as the maximum amount. Although this measure seems rather drastic, its long-term impact will nevertheless be rather limited because limitations in world supply will already present themselves quite soon. The above measure would cause European long-term demand to decline some 15 per cent only vis-à-vis its base projection, and world demand at the international market some 7 per cent. One therefore should not be surprised that the above measure turned out to have no more than a negligible impact on the Indonesian timber industry, the more so if one also recognizes that, mainly due to transportation costs, the Indonesian exports to Europe are relatively minor.

Joint quota by the developed countries
To what extent will the above picture change if all developed countries start to reduce their timber imports? Suppose the OECD countries together with Central and Eastern Europe collectively decided to limit their imports to the 1987 level. In the short term, this results in a global demand some 5 per cent lower than under the base case assumptions during the 1990s. Indonesia's cumulative exports to Japan will be some 30 per cent less during the 1990s, total cumulative exports some 7 per cent.

Paradoxically enough, such a policy would favour Indonesia's long-term (i.e. after 2010) market position. The main explanation is that due to less international demand, commercial reserves which are mainly responsible for future restrictions in supply, are less reduced. Precisely because the reduction in the amount of commercial reserves is Indonesia's relatively weak spot, such a policy supports Indonesia's potential as a future supplier. More specifically, supply restrictions to satisfy domestic demand will, after 2010, be some 5 per cent less severe, and total Indonesian exports will not even decline although there will be a shift away from exports to Japan towards the NICs (unless they started to join the quota). Indeed, the NICs might develop into major future consumers of Indonesian tropical hardwood.[2]

An Indonesian export quota

In this scenario there is only room for exports from Indonesia if domestic demand is less then some 130 million m³ per annum; if domestic demand surpasses this level, however, supply will adjust but exports will disappear. The model outcomes show that in such a case Indonesia will cease to export from the middle of the next decade (i.e. after 2000). It is capable of meeting domestic demand much better, however. Indeed, long-term supply which is available for the domestic processing industries will be some 50 per cent larger than otherwise. This is an important finding in view of the above mentioned official projections of the Indonesian government.

Obviously this measure would reduce the Indonesian exports. To illustrate, during the period up to 2005 the cumulative exports to Japan would be 80 per cent less than in the base case; other countries would obviously fill this gap, so that at the global scale commercial reserves would hardly be preserved. All the above policies do not or at most hardly affect deforestation processes,[3] the main factor responsible for forest degradation.

Population policy

If the Indonesian government succeeded in reducing its annual population growth rate during the simulation period from 1.6 per cent to 1 per cent, the following consequences for both deforestation and logging would result:

1) Less population growth would cause demand for farmland to raise by 1.4 per cent annually against 2 per cent in the base case. This causes cumulative deforestation to be almost half the size of what it by 2030 otherwise would have been: 27 million ha instead of 45 million ha!
2) The amount available to meet domestic demand is also favourably affected: in the long term this will be some 10 per cent more.

---

[2] In a simulation wherein the NICs would also join the OECD quota there will be a likely shift from the processing activities in the NICs back to the producing regions. Indonesia's market position at these markets will, however, remain strong because of the relatively low transportation costs.

[3] The model does not yet take the direct relationship between logging and deforestation into account, the main factor responsible for forest degradation.

Preservation through nature reserves

Declaring parts of tropical rainforests as nature reserves (whether or not stimulated through, e.g., the donor community by way of debt-for-nature swaps) could be viewed as a serious policy option indeed, at least if problems with respect to safeguarding and control can be overcome. IDIOM basically distinguishes between three sorts of tropical areas: the part containing the commercial reserves; the inaccessible and thus unproductive area; and the area in recovery from former logging. Thus one faces the issue of how an increase in the national reservations should be assigned to the three types of forest. Here the assumption is made that reservations will be created in the first two categories in proportion to their respective surface areas, but other assumptions might also be viable.

Suppose 10 per cent of the tropical rainforest area in Indonesia, some 12 million ha, is declared a national nature reserve. What would the implication be in terms of exports, domestic consumption, and deforestation? The simulations show the impact to be rather modest: the cumulative exports will decrease by some 2 per cent over the whole simulation period; also there will be only 1 per cent less timber available for domestic purposes than otherwise, and cumulative deforestation will total no more than about 0.2 per cent less than in the base scenario. It looks like the impact of this quite drastic policy option is rather disappointing. The reason is basically that one just moves into the non-protected areas earlier.

Deforestation and agricultural productivity

For a complete assessment of all policy options it can be worthwhile to focus precisely on another of these often neglected indirect relations. A simulation was therefore carried out linking agricultural development with deforestation. The assumption was that through intensive policy action the yearly increase of agricultural productivity would raise 1 percentage point throughout the simulation period, roughly from some 3 per cent to some 4 per cent. Although such a jump in productivity requires a massive effort, historical data (also with respect to other regions) show that it has been possible to realize such an achievement in the past.

The simulation results make clear that the impact of such a change would be enormous! Cumulative deforestation would be over 50 per cent less; commercial reserves would end up some 25 per cent larger than otherwise; there would be some 7 per cent more resources available to meet the domestic demand for timber; and the annual Indonesian tropical timber exports could expand to a level on average some 6 per cent above the base level. Because of this, policy options along these lines definitely merit further scrutiny.

CONCLUSION

A modelling exercise was carried out to provide some projections on production and trade of tropical timber in the various regions of the world and Indonesia in particular, and consequently on degradation processes of tropical forest. The main conclusion is that within a relatively short period of time there will be a shortage of timber supply on a global scale, and that the present rate of deforestation will continue unless drastic policy steps are taken. If not, by 2030 some 20 per cent of the tropical forests, and some 80 per cent of the commercial timber reserves will have disappeared. Moreover the volume of total international trade in roundwood equivalents in tropical timber (products) will be 70 per cent less than at the current time.

By narrowing down the analysis to the Indonesian case, it was shown that Indonesia will suffer from this phenomena even more than most of the other producing regions, due to the fact that a relatively large share of its commercial reserves have already been exploited. Consequently the country will soon loose its dominant position in the international market, and face shortages in supply that may even conflict with recent domestic economic planning.

A promising factor, however, is Indonesia's front position in discussing policy options to deal with this worldwide environmental issue. As a preliminary exercise here, some rough policy options have been simulated consecutively; the main goal was to gain an initial impression of the orders of magnitude of their impact. The picture emerging was that limiting international timber trade can only work if carried out in a coordinated way, but even then will mainly relocate rather than reduce demand. The reduction of deforestation is also much less than many of us would have expected. A similar, rather disappointing impression came out of the options to cut down only Indonesian timber exports or expand the size of national nature reserves in the same country considerably.

It looks like a serious impact on tropical forest degradation can only be expected from a comprehensive coordinated approach, including traditional developmental policy options such as reducing the population pressure, creating buffer zones near the tropical forests, and raising agricultural productivity. Massive and integrated efforts along these lines indeed seem capable of substantially mitigating the trends towards progressive tropical forest degradation. They therefore deserve to be further evaluated as instruments of environmental policy in a broad international framework, along with the timber trade and production oriented policy devices.

APPENDIX 3

EXPORTS OF TROPICAL TIMBER
BY MAIN EXPORTING COUNTRIES
(in 1000 m³ unless indicated otherwise)

	Export value ($m)	Round-wood	Sawn-wood	Ply-wood	Veneer	Total (1000 m³ re)
Malaysia	1977	20,526	3204	498	410	28,280
Indonesia	1740	19	2525	4626	79	15,410
Philippines	227	653	547	262	61	2370
Singapore	369	20	803	632	43	3020
Ivory Coast	240	1010	469	23	75	2060
Cameroon	77	533	87	3	20	740
Gabon	120	1036	2	46	7	1160
Brazil	221	15	419	224	56	1400

Source: Bos en Hout Berichten, 1988, no. 11.

APPENDIX 4

EXPORT SHARES IN TOTAL TIMBER PRODUCTION (%)

	1985	1990	2000	2025
Asia	36	65	47	26
Africa	24	8	8	40
South America	3	27	44	34

Sources: NEI (1989), Table 2.5 and own projections.

APPENDIX 5

SARUM: REGIONAL DISAGGREGATION

North America	Europe	Developed Pacific	Eastern Europe & USSR	Tropical Latin America	Tropical Africa	Tropical Asia	Other Latin America	Other Africa	Other Asia
Canada	Austria	Japan	Albania	Bolivia	Angola	Burma	Antigua	Botswana	Afghanistan
United States	Belgium	Australia	Bulgaria	Brazil	Benin	Indonesia	Argentina	Burkino Faso	Bangladesh
	Denmark	New Zealand	Czechoslov.	Colombia	Cameroon	Laos	Bahamas	Burundi	Bhutan
	Finland	American Samoa	Hungary	Ecuador	Centr. Af. Rep.	Malaysia	Barbados	Cape Verde	Brunei
	France	Cook Isl.	Poland	Guyana	Congo	Papua New Guinea	Belize	Chad	China
	Germany	Fiji	Romania	Peru	Cote d'Ivoire	Philippines	Bermuda	Comoros	India
	Greece	Fr. Polynesia	USSR	Suriname	Gabon	Thailand	Br. Virgin Isl.	Djibouti	Kampuchea
	Iceland	Guam	Yugoslavia	Venezuela	Ghana	Vietnam	Cayman Isl.	Eq. Guinea	Maldives
	Ireland	Kiribati			Guinea		Cuba	Ethiopia	Mongolia
	Italy	Nauru			Kenya		Chile	Gambia	Nepal
	Luxembourg	New Caledonia			Madagascar		Costa Rica	Guinea-Buissau	Pakistan
	Netherlands	Niue			Nigeria		Dominica	Lesotho	Sri Lanka
	Norway	Pacific Isl.			Senegal		Dominican Rep.	Liberia	Hong Kong
	Portugal	Samoa			Sierra Leone		El Salvador	Malawi	Korea Dem. Rep.
	Spain	Solomon Isl.			Tanzania		Falklands	Mali	Korea Rep.
	Sweden	Tokelau			Uganda		Fr. Guyana	Mauritania	Macau
	Switzerland	Tonga			Zaire		Greenland	Mauritius	Singapore
	Un. Kingdom	Tuvalu					Grenada	Mozambique	Taiwan
	Israel	Vanuatu					Guadeloupe	Namibia	Bahrain
	Malta	Wallis							Iran

North America	Europe	Developed Pacific	Eastern Europe & USSR	Tropical Latin America	Tropical Africa	Tropical Asia	Other Latin America	Other Africa	Other Asia
							Guatemala	Niger	Iraq
							Haiti	Reunion	Jordon
							Honduras	Rwanda	Kuwait
							Jamaica	Sao Tome	Lebanon
							Martinique	Seychelles	Oman
							Mexico	Somalia	Qatar
							Montserrat	St. Helena	Saudi Arabia
							Neth. Antilles	Swaziland	Syrian
							Nicaragua	Togo	Turkey
							Panama	Zambia	Un. Arab Emirates
							Paraguay	Zimbabwe	Yemen Dem.
							Puerto Rico	Sudan	Yemen
							St. Christopher	South Africa	
							St. Lucia	Algeria	
							St. Pierre	Egypt	
							St. Vinc.	Libya	
							St. Kitts	Morocco	
							Trinidad	Tunisia	
							Turks. Caicos Isl.		
							Uruguay		
							U.S. Virgin Isl.		

APPENDIX 6

LIST OF BACKGROUND DOCUMENTS

Blom, M., *The future role of the tropical rain forests: an economic outlook for Indonesia*, IDE paper 9101, May 1991.

Blom, M., J.A.M. Hoogeveen and A. van der Linden, *TROPFORM: a model of the tropical rain forest, an analysis*, IDE paper 9005, April 1990.

Bruyn, R. de, *The tropical rainforest and the timber trade: enemies or friends?*, IDE paper 9202, January 1992.

Dulleman M., *Deforestation in Brazil's Amazon region: causes, consequences, and public policy*, IDE paper 9301, January 1993.

Hoogeveen, J.A.M., *Bosbouw- en bosbeheermodellen: een eerste literatuurverkenning* (Forestry and forest conservation models: a literature survey), IDE paper 8915, December 1989.

Hoogeveen, J.A.M., *Agricultural productivity in the tropics*, IDE paper 9102, June 1991.

Hoogeveen, J.A.M., *Agricultural productivity in the humid tropics: some scenarios*, IDE paper 9201, February 1992.

Jepma, C.J. and M. Blom, *Global trends in tropical forest degradation: the Indonesian case, policy simulations with the help of IDIOM*, IDE paper 9007, October 1990. Also published in: W.J.M. Heijman and J.J. Krabbe (eds.), Wageningen Economic Studies no. 24, *Issues of environmental economic policy*, Wageningen, 1992, pp.185–214.

Parker, K.T., *Migration models and SARUM/TROPFORM*, IDE paper 9107, December 1991.

Schippers, M., *Land use in the tropics, Agroforestry as a solution to deforestation?*, IDE paper 9203, July 1992.

BIBLIOGRAPHY

African Business (1992), Timber companies in Ghana 'make off with donor funds', no.165, May, p.36.

Albert B. (1992), Indian lands, environmental policy and military geopolitics in the development of the Brazilian Amazon: the case of the Yanomami, *Development and Change*, vol.23, pp.35-70.

Allen J.C. (1985), Wood energy and preservation of woodlands in semi-arid developing countries, the case of the Dodoma region Tanzania, *Journal of Development Economics*, vol.19, pp.59-84.

Alriksson B., Ohlsson A. (1990), *Farming systems with special reference to agroforestry, a literature review and a field study in Babati district, Tanzania*, SUAS working paper 129, Uppsala.

Amelung, T., Diehl M. (1992), *Deforestation of tropical rainforests: economic causes and impact on development*, Tubingen, Mohr.

Anderson A.B. (1990a), Deforestation in Amazonia: Dynamics, causes, and alternatives, in Anderson, A.B. (ed.), *Alternatives to deforestation: Steps toward sustainable use of the Amazon rain forest*, Columbia University Press, New York.

Anderson A.B. (1990b), Smokestacks in the rainforest: industrial development and deforestation in the Amazon Basin, *World Development*, vol.18, pp.1191-1205.

Anderson D., Fishwick R. (1984), *Fuelwood consumption and deforestation in African countries*, World Bank Staff Working Papers no.704, Washington D.C.

Ark B. van (1991), *International comparisons of agricultural productivity*, note prepared for and presented at a workshop on deforestation in the humid tropics, University of Groningen, December 6.

Armitage I., Kuswanda M. (1989), Forest Management for sustainable production and conservation in Indonesia, *Indonesia Forestry Studies*, Field Document no. I-2, Directorate General of Forest Utilization, Ministry of Forestry/FAO, Jakarta, October.

Armitage J., Schramm G. (1989), Managing the supply of and demand for fuelwood in Africa, in: Schramm G., Warford J.J. (eds), *Environmental management and economic development*, World Bank, Washington D.C., pp.139-171.

Arnold J.E.M. (1983), Economic considerations in agroforestry projects, *Agroforestry Systems*, no.1, pp.299-311.

Arnold J.E.M. (1987), Economic considerations in agroforestry projects, in Steppler, H.A., Nair P.K.R. (eds), *Agroforestry, a decade of development*, ICRAF, Nairobi.

Asian Development Bank (1989), *Sector paper on forestry*, July.

Baer W. (1983), *The Brazilian economy: growth and development*, Praeger Publishing, New York.

Barbier et al. (1992), *The economic linkages between the*

international trade in tropical timber and the sustainable management of tropical forests, draft main report and annexes A-K, London Environmental Economics Centre, London UK.

Bare B.B. et al. (1984), A survey of systems analysis models in forestry and the forest products industries, *European Journal of Operational Research*, vol.18, no.1, pp.1-18.

Barnes D.F. (1990), *Population growth, wood fuels, and resource problems in sub-Saharan Africa*, Energy Series Paper no.26, The World Bank Industry and Energy Department, Washington D.C., March.

Barraclough S., Ghimire K. (1990), *The social dynamics of deforestation in developing countries: principal issues and research priorities*, UNRISD discussion paper 16, Geneva, November.

Barrett S. (1991), Optimal soil conservation and the reform of agricultural pricing policies, *Journal of Development Economics*, vol.36, no.2, pp.167-187, October.

Bass S., Poore D., Romijn B. (1992), *ITTO and the future in relation to sustainable development*, Report prepared by AIDenvironment and the International Institute for Environment and Development, September.

Baumol W.J., Oates W.E. (1988), *The theory of environmental policy*, Cambridge University Press, Cambridge.

Beer De J., McDermott M. (1989), The economic value of non-timber forest products in Southeast Asia, IUCN, Amsterdam.

Bilsborrow R.E., Okoth Ogendo H.W.O. (1992), Population-driven changes in land-use in developing countries, *AMBIO*, vol.21, no.1, February.

Binswanger H.P. (1991), Brazilian policies that encourage deforestation in the Amazon, *World Development*, vol.19, no.7, pp.821-829.

Birgegard L.E. (1991), Forestry activities are not the answer to deforestation, *Forests, Trees and People Newsletter*, no.14, pp.35-37, October.

Bishop, J.T., Bann C.A. (1992), *Willingness to pay for sustainably produced timber*.

Blom M. (1991), *The future role of the tropical rainforests: an economic outlook for Indonesia*, IDE paper 9101, Groningen, May.

Blom M., Hoogeveen J.A.M., Linden A. van der (1990), *TROPFORM: a model of the tropical rain forests, an analysis*, IDE paper 9005, Groningen, April.

Bongaarts J., Mauldin W.P., Phillips F. (1990), The demographic impact of family planning programs, *Stud. Fam. Plan.*, vol.21, no.6.

B.O.S. (1991), *Newsletter*, Wageningen, August.

Boserup E. (1965), *The conditions of agricultural growth: the economics of agrarian change under population pressure*, Allen and Unwin, London.

Boserup E. (1981), *Population and technology*, Basic Blackwell, Oxford.

Boulier B.L. (1985), *Evaluating unmet need for contraception, estimates for thirty-six developing countries*, World Bank Staff Working Papers no.678, Population and Development Series no.3, World Bank, Washington D.C.

Bourke I.J. (1991), Domestic timber markets: important outlets for the developing countries, *Unasylva*, vol.42, no.167, pp.16-24.

Braat L.C., van Lierop W.J.F. (eds) (1987), *Economic-ecological modelling*, IIASA-studies in regional science and urban economics no.16, Elsevier Science Publishers, Amsterdam.

Bruyn R. de (1992), *The tropical rainforest and the timber trade: enemies or friends?*, M.A. thesis, Faculty of Economics, University of Groningen, January.

Buongiorno, J., Manarung T. (1992), *Predicted effects of an import tax in the European community on the international trade in tropical timbers*, mimeo., Department of Forestry, University of Wisconsin, Madison, USA.

Cardoso E.A. (1992), Deficit finance and monetary dynamics in Brazil and Mexico, *Journal of Development Economics*, vol.37, pp.173-197.

Cavalcanti C. (1991), Government policy and ecological concerns: Some lessons from the Brazilian experience, in Constanza R. (ed.), *Ecological Economics*, Columbia University Press, New York.

Clarke H.R., Reed W.J. (1989), The tree-cutting problem in a stochastic environment, the case of age dependent growth, *Journal of Economic Dynamics and Control*, vol.13, no.4, pp.569-595.

Cohen, B.I. (1986), Relative effects of foreign capital and larger exports on economic development, *Review of Economics and Statistics*, no.9, pp.281-284.

Conservation International (1989), *The debt-for-nature exchange, a tool for international conservation*, CI, Washington D.C.

Constantino, L.F. (1988), Analysis of the international and domestic demand for Indonesian wood products, mimeo., *Report for the FAO*, Department of Rural Economy, University of Alberta, Edmonton, Canada.

Cox P.A., Elmqvist T. (1991), Indigenous control of tropical rainforest reserves: an alternative strategy for conservation, *Ambio*, vol.20, no.7, pp.317-321, November.

Crabbé P.J., Long N. van, (1989), Optimal forest rotation under monopoly and competition, *Journal of Environmental Economics and Management*, vol.17, no.1, pp.54-65.

Cross M. (1988), Spare the tree and spoil the forest, *New Scientist*, no.26, November.

Dasgupta P.(1992), Population, resources and poverty, *Ambio*, vol.21, no.1, February.

Dennis, T.L., Dennis L.B. (1991), *Management Science*, West, St. Paul.

Dewees P. (1989), The woodfuel crisis reconsidered: observations on the dynamics of abundance and scarcity, *World Development*, vol.17, no. 8.

Doolette, J.B., Smyle J.W. (1990), Soil and moisture conservation technologies: review of literature, in Doolette, J.B., Magrath W.B.

(eds), *Watershed development in Asia: strategies and technologies*, World Bank Technical Paper 127, Washington D.C.

Dorland S. et al. (1988), *Shifting cultivation nu, multistorey farming straks?* (Shifting cultivation now, multistorey farming later?), Projectgroep Shifting Cultivation, LU Wageningen.

Dulleman M. (1993), *Deforestation in Brazil's Amazon region: causes, consequences, and public policy*, M.A. thesis, University of Groningen, paper 9301, January.

ECE/FAO (1986), *European timber trends and prospects to the year 2000 and beyond*, vol.1 and vol.2, New York.

Environmental Strategies Europe (ESE) (1992), *Identification and Analysis of measures and instruments concerning trade in tropical wood*, Vol.I–IV, Report submitted to the Commission of the European Communities, September.

FAO, *Yearbook of forest products*, various volumes, Rome.

FAO (1985), *Intercountry comparisons of agricultural production aggregates*, FAO Economic and Social Development Paper 61, FAO Statistics Division, Rome.

FAO (1987), *Small-scale forest-based processing enterprises*, FAO Forestry Paper 79, Rome.

FAO (1988), *An interim report on the state of forest resources in the developing countries*, Rome.

FAO (1989), *Forestry and food security*, FAO Forestry Paper 90, Rome.

FAO (1991), *Yearbook of production and trade 1990*, vol.44, Rome.

FAO (1993), *Forest resources assessment 1990, tropical countries*, FAO Forestry Paper 112, Rome.

FAO/GOI (1990), *Fuelwood and charcoal, situation and outlook of the forestry sector of Indonesia*, vol.3, Forest Resource Utilization, Jakarta.

Fearnside P.M. (1990), Predominant land uses in Brazilian Amazonia, in Anderson A.B. (ed.), *Alternatives to deforestation: steps toward sustainable use of the Amazon rain forest*, Columbia University Press, New York.

Feldstein M. (1980), Inflation, tax rules and the prices of land and gold, *Journal of Public Economics*, vol.14, pp.309–317.

Ferguson-Bisson (1992), Rational land management in the face of demographic pressure: obstacles and opportunities for rural men and women, *AMBIO*, vol.21, no.1, February.

Ferguson I.S., Muñoz-Reyes Navarro J. (1992), *Working document for ITTO expert panel on resources needed by producer countries to achieve sustainable management by the year 2000*, ITTO, Yokohama.

Gane M. (1986), TIMPLAN: a planning system for industrial timber-based development, *Commonwealth Forestry Review*, vol.65, no.1, pp.41–49.

Geertz C. (1963), *Agricultural involution, the process of ecological change in Indonesia*, the University of California, Berkeley.

Gigengack A.R., Jepma C.J., Lanjouw G.J., Schweigman C. (1985), *The use of a world model for the analysis of North-South interdependence and problems of development and security*, Development and Security Series, Groningen.

Gillis M., Perkins D.H., Roemer M., Snodgrass D.R. (1987), *Economics of development*, W.W. Norton and Company, New York.

Grainger A. (1986), *The future of the tropical rain forests in the world forest economy*, Ph.D. Thesis, University of Oxford.

Grainger A. (1987), TROPFORM: a model of future hardwood supplies, *Proceedings of the CINTRAFOR symposium on forest sector and trade models*, University of Washington, Seattle.

Hadley M., Schreckenberg K. (1989), *Contributing to sustained resource use in the humid and sub-humid tropics: some research approaches and insights*, MAB Digest 3, Unesco, Paris.

Hagget P. (1979), *Geography, a modern synthesis*, third edition, Harper & Row, New York.

Harris J., Todaro M.P. (1970), Migration, unemployment and development: a two-sector analysis, *American Economic Review*, no.60, pp.126–142.

Harrison P. (1991), Beyond the blame-game: population-environment links, *Populi*, vol.17, no.3.

Hassan R.M., Hertzler G. (1988), Deforestation from the overexploitation of wood resources as a cooking fuel, a dynamic approach to pricing energy resources in Sudan, *Energy Economics*, vol.10, no.2, pp.163–168.

Hayami Y., Ruttan V.W. (1985), *Agricultural development, an international perspective*, The Johns Hopkins University Press, Baltimore.

Hecht S.B. (1985), Environment, development and politics: capital accumulation and the livestock sector in Eastern Amazonia, *World Development*, vol.13, pp.663–684.

Hill K. (1992), Fertility and mortality trends in the developing world, *AMBIO*, vol.21, no.1, February.

Hinrichsen D. (1991), The need to balance population with resources: four case studies, *Populi*, vol.18, no.3.

Hoogeveen J.A.M. (1989), *Bosbouw- en bosbeheermodellen: een eerste literatuurverkenning* (Forestry and forest conservation models: a literature survey), paper 8915, Groningen, December.

Hoogeveen J.A.M. (1991), *Agricultural productivity in the tropics*, paper 9102, Groningen, June.

Hoogeveen J.A.M. (1992), *Agricultural productivity in the humid tropics: some scenarios*, paper 9201, Groningen, February.

Hoogeveen J.A.M., Blom M. (1989), *The SARUM dataset as used for estimating future financial requirements of developing regions*, paper 8910, August.

Howe C.W. (1987), *Natural resource economics, issues, analysis and policy*, John Wiley & Sons, New York.

IBGE (1990), *Anuàrio estatìstico do Brasil 1990*, Rio de Janeiro.

ICRAF (1991), *Annual report 1990*, Nairobi.

IMF (1992), *Direction of trade statistics yearbook*, Washington D.C.

Imhoff E. van (1991), De Club van Rome over het bevolkingsvraagstuk

(The 'Rome Club' on the population issue), *Demos*, vol.7, no.10.

Ingram, C.D., L.F. Constantino, M. Mansyur (1989), Statistical information related to the Indonesian forestry sector, in *Indonesia Forestry Studies*, working paper no. 5, Ministry of Forestry, Government of Indonesia, Food and Agriculture Organization of the United Nations, Jakarta.

IOV (1990), *Hulp of handel* (Aid or trade), Inspectie Ontwikkelingssamenwerking te Velde (Field Inspection of Development Cooperation), Ministry of Foreign Affairs, The Hague.

IUSSP (1980), *Projecting migration for integrated rural and urban development planning*, IUSSP Papers, no.16, Liege (Belgium).

Jacobs M. (1988), *The tropical rain forest, a first encounter*, Springer Verlag, Berlin, Heidelberg.

Janssen W.G., Sanint L.R. (1991), Economic trends in Latin America: roles for agriculture and new technology, *Food Policy*, vol.16, pp.474–485.

Jepma C.J., Blom M. (1990), *Global trends in tropical forest degradation: the Indonesian case, policy simulations with the help of IDIOM*, paper 9007, Groningen, October.

Jepma, C.J. (1992), *LDC financial requirements*, Avebury, Aldershot.

Johansson P.O., Löfgren K.G. (1985), *The economics of forestry & natural resources*, Basil Blackwell, Oxford, U.K.

Johnson H.G. (1967), *Economic policies towards less developed countries*, London.

Kannappan S. (1985), Urban employment and the labor market in developing nations, *Economic Development and Cultural Change*, vol.33, pp.699–730.

Kerkhof P. (1990), *Agroforestry in Africa, a survey of project experience*, Panos, London.

Kleinpenning J.M.G. (1991), The environment in developing countries: an abiding source of concern, *Internationale Spectator*, xlv-11, pp.706–709, November.

Kula E. (1988), *The economics of forestry, modern theory and practice*, Croom Helm, London.

Kyle S.C., Cunha A.S. (1992), National factor markets and the macroeconomic context for environmental destruction in the Brazilian Amazon, *Development and Change*, vol.23, pp.7–33.

Lagemann (1977), *Traditional African farming systems in eastern Nigeria: an analysis of reaction to increasing population pressure*, Africa Studien, Weltforum Verlag, Munich.

LEEC (1993), *The economic linkages between the international trade in tropical timber and the sustainable management of tropical forests*, draft report for ITTO, London Environmental Economics Centre, International Institute for Environment and Development, London.

Levi J., Havinden M. (1982), *Economics of African agriculture*, Longman, Harlow, UK.

Lewis A. (1954), *Economic development with unlimited supplies of labour*, Manchester School.

Livingstone, I. (1990), Population growth and rural labor absorption in Eastern and Southern Africa, in McNicoll G., Cain M. (eds), *Rural Development and Population, Institutions and Policy*, New York.

Longman K.A., Jenik J. (1974), *Tropical forests and its environment*.

Lundgren B., Raintree J.B. (1983), Sustained agroforestry, in *Agricultural research for development: potentials and challenges in Asia*, report of a conference held October 24-29 1982, Indonesia, International Service for National Agricultural Research (ISNAR), The Hague.

Lyons, M. (1990), *The timber giant, Indonesia: country upgrade*, Asia Pacific Forest Industries.

Mahar D.J. (1979), *Frontier development policy in Brazil: a study of Amazonia*, Praeger Publishers, New York.

Mahar D.J. (1989), Deforestation in Brazil's Amazon region: magnitude, rate, and causes, in Schramm G., Warford J.J. (eds), *Environmental management and economic development*, The Johns Hopkins University Press, Baltimore.

Mahar J.H. (ed.) (1985), *Rapid population growth and human carrying capacity, two perspectives*, World Bank Staff Working Papers no.690, Population and Development Series no.15, World Bank, Washington D.C.

Manolas F. (1991), *Tropical rainforests: sustainability or destruction?, a soft systems approach to the sustainability of rainforests*, M.A. thesis, University of Kent, Canterbury.

Manshard W., Morgan W.B. (1988), Introduction, in Manshard W., Morgan W.B. (eds), *Agricultural expansion and pioneer settlements in the humid tropics*, the United Nations University, Tokyo, pp.1-6.

McCleary R.M. (1991), The international community's claim to rights in Brazilian Amazonia, *Political Studies*, vol.39, pp.691-707.

McKillop W., Wibe S. (1987), Demand for sawnwood and panels, in Markku Kallio, Dijkstra D.P., Binkely C.S. (eds), *The global forest sector, an analytical perspective*, IIASA, Laxenburg.

McNicoll G., Cain M. (eds) (1990), *Rural development and population, institutions and policy*, New York.

Meadows D.H. et al. (1972), *The limits to growth*, New York.

Millikan, *The dialectics of devastation: tropical forest deforestation, land degradation and society in Rondonia, Brazil*, Master's Thesis, University of California, Berkeley.

Mors M. (1991), *The economics of policies to stabilize or reduce greenhouse gas emissions: the case of CO_2*, Economic Papers no.87, Commision of the European Communities, Brussels.

Mwandosya M.J., Luhanga M.L. (1985), An analytical model for a biomass system, *Energy*, vol.10, no.9, pp.1023-1028.

Myers N. (1984), *The primary source, tropical forests and our future*, W.W. Norton & Company, New York.

Myers N. (1989), *Deforestation rates in tropical forests and their climatic implications*, Friends of the Earth, London.

Myers N. (1991), Tropical forests: present status and future outlook, *Climatic Change*, vol.19, pp.3-32, September.

Myers N. (1992), Population/environment linkages: discontinuities ahead, *AMBIO*, vol.21, no.1, February.

Nair P.K.R. (1990), *The prospects for agroforestry in the tropics*, World Bank Technical Paper no.131, Washington D.C.

Nair C.T.S., Krishnankutty C.N. (1984), Socio-economic factors influencing farm forestry: a case study of tree cropping in the homesteads in Kerala, India, in *Community Forestry: socio-economic aspects*, Bangkok: FAO/East-West Center.

Nationale Advies Raad voor Ontwikkelingssamenwerking (National Advisory Council for Development Cooperation) (1990), *Voorlopig regeringsstandpunt tropisch regenwoud* (Provisional standpoint of the government on the tropical rainforest), summary, April.

Netherlands Economic Institute (1989), *An import surcharge on the import of tropical timber in the European Community: an evaluation*, NEI, Rotterdam.

OECD (1979), *Facing the future*, OECD Interfutures, Paris.

OECD (1991a), *Financing and external debt of developing countries*, 1990 Survey, Paris.

OECD (1991b), *Development co-operation*, Development Assistance Committee, OECD, Paris.

Otto (1990), Plattelandsontwikkeling en herbebossing, voorwaarden tot behoud van tropisch regenwoud (Rural development and reforestation, conditions for the conservation of the tropical rainforest), *Bos en Hout berichten* no.3, Bos en Hout Foundation, Wageningen.

Parker, K.T. (1979), Modelling inter-regional activity by means of trade biases, *Proceedings of the IFAC/IFORS conference on dynamic modelling*, North Holland.

Parker K.T. (1991), *Migration models and SARUM/TROPFORM*, paper 9106, Groningen.

Parker K.T. (1992), *Labour migration and the modelling of deforestation in an inter-sectoral and global context*, Working Paper No.15, Canterbury Business School, University of Kent, May.

Pearce F. (1992), First aid for the Amazon, *New Scientist*, vol.133, no.1814, pp.42–46, March.

Pearce P., Barbier E., Markandya A. (1990), *Sustainable development, economics and environment in the Third World*, Edward Elgar Publishing Limited, Aldershot.

Pereira L.B. (1990), Brazil's inflation and the Cruzado Plan, 1985-1988, in Falk P.S. (ed.), *Inflation: are we next?; hyperinflation and solutions in Argentina, Brazil, and Israel*, Lynne Rienner Publishers, Boulder & London.

Pessino C. (1991), Sequential migration theory and evidence from Peru, *Journal of Development Economics*, vol.36.

Peters C. et al. (1989), *Nature Magazine*, June 29.

Pingali P.L. (1990), Institutional and environmental constraints to agricultural intensification, in McNicoll G., Cain M. (eds), *Rural development and populations: institutions and policy*, Oxford University Press (USA), pp.243–261.

Poldy F. et al. (1986), *The Area Model handbook*, Australian Resources and Environmental Assessment (AREA) project, Canberra.

Poore D. et al. (1989), *No timber without trees, sustainability in the tropical forest*, Earthscan Publications, London.

Raintree J.B. (1983), Strategies for enhancing the adoptability of agroforestry innovations, *Agroforestry Systems*, no.1, pp.173-187.

Raintree J.B. (1991), *Socio-economic attributes of trees and tree planting practices*, FAO Community Forestry Note no.9, Rome.

Raintree J.B., Warner K. (1986), Agroforestry pathways for the intensification of shifting cultivation, *Agroforestry Systems*, no.4, pp.39-54, Martinus Nijhoff/Dr W. Junk Publishers, Dordrecht.

Ramirez A., Seré C., Uquillas J., An economic analysis of improved agroforestry practices in the Amazon lowlands of Ecuador, *Agroforestry Systems*, no. 17, pp.65-86.

Regeringsstandpunt tropisch regenwoud (The government's standpoint on tropical rainforests) (1991), the Hague.

Reijntjes C., Haverkort B., Waters-Bayer A. (1992), *Farming for the future, an introduction to low-external-input and sustainable agriculture*, The Macmillan Press Ltd, London.

Renaud B. (1979), *National urbanization policies in developing countries*, World Bank Staff Working Paper, no.347.

Repetto R. (1988), *The forest for the trees? Government policies and the misuse of forest resources*, WRI, Washington D.C.

Repetto R. (1989), Economic incentives for sustainable production in Schramm G., Warford J.J. (eds), *Environmental management and economic development*, World Bank, Washington D.C., pp.69-86.

Repetto R. (1990), Deforestation in the tropics, *Scientific American*, vol.262, no.4, pp.18-24, April.

Robinson N.A., Hassan P., Burhenne-Guilmin F. (1992), (for the Commission on Environmental Law of the World Conservation Union and the International Union for the Conservation of Nature and Natural Resources (IUCN)), *Agenda 21 & the UNCED Proceedings*, Vol.4, Oceana Publications, New York.

Ruthenberg H. (1980), *Farming systems in the tropics*, third edition, Oxford University Press.

Ryan J.C. (1991), Goods from the woods, *Forests, Trees and People*, Newsletter no.14, pp.23-30, October.

Sanchez P.A. (1987), Soil productivity and sustainability in agroforestry systems, in Steppler H.A. and Nair P.K.R. (eds), *Agroforestry: a decade of development*, ICRAF, Nairobi.

Santiago C.E., Thorbecke E. (1988), A multisectoral framework for analysis of labor mobility and development in LDCs: an application to postwar Puerto Rico, *Economic Development and Cultural Change*, no.37, pp.127-148.

Sawyer D. (1990), The future of deforestation in Amazonia: a socioeconomic and political analysis, in Anderson A.B.(ed.), *Alternatives to deforestation: steps toward sustainable use of the Amazon rain forest*, Columbia University Press, New York.

Schippers M. (1992), *Land use in the tropics, agroforestry as a solution to deforestation?*, IDE paper 9203, July, Groningen.

Schmidt R., *Sustainable management of tropical moist forests*, paper presented at the ASEAN sub-regional Seminar, Indonesia.

Schoenmakers J. (1992), Tropisch hout wordt minder belangrijk (Tropical timber becomes less important), *Financieel Economisch Magazine*, vol. 23, no.10, pp.18-20, May.

Secretariat General of Ministry of Forestry (1990), *Statistik Kehutanan Indonesia* (Forestry Statistics of Indonesia), 1988-1989, Biro Perencanaan (Bureau of Planning).

Sedjo R.A., Clawson M. (1984), Global forests, in Simon J.L. and Khan H. (eds), *The Resourceful Earth*, Basil Blackwell, Oxford.

Serrão A.D., Toledo J.M. (1990), The search for sustainability in Amazonian pastures, in Anderson A.B. (ed.), *Alternatives to deforestation: steps toward sustainable use of the Amazon rain forest*, Columbia University Press, New York.

Sewastynowicz J. (1986), Two step migration and upward mobility on the frontier: the safety valve effect in Pejibaye, Costa Rica, *Economic Development and Cultural Change*, vol.34, pp.731-753.

Shane D.R. (1986), *Hoofprints on the forest, cattle ranching and the destruction of Latin America's tropical forests*, Philadelphia.

Sharma N., Rowe R. (1992), Managing the world's forests, *Finance and Development*, vol.29, no.2, pp.31-33, June.

Shields G.M., Shields M.P. (1989), Family migration and nonmarket activities in Costa Rica, *Economic Development and Cultural Change*, vol.37, pp.73-88.

Shukla V., Stark O. (1986), Urban external economies and optimal migration, *Research in Human Capital and Development: Migration, Human Capital and Development*, vol.4, pp.139-146.

Standing G. (1984), *Population mobility and productive relations, demographic links and policy evolution*, World Bank Staff Working Papers no.695, Population and Development Series no.20, World Bank, Washington D.C.

Staveren I. van (1992), De malthusiaanse valkuil (The Malthusian trap), Chapter 4, in *Over bevolking, een analyse van het denken over bevolkingsgroei en de praktijk van de bevolkingspolitiek* (On population, an analysis of the view on population growth and the practice of population policy), Osaci Foundation, Utrecht.

Stichting Bos en Hout (1991), *Market intelligence: analysis of the wood flow as a basis for an early warning system for the tropical timber market*, SBH, Wageningen.

Strong M.F. (1991), The challenge before the earth summit, *Populi*, vol.18, no.3.

Sutter H. (1989), Forest resources and land use in Indonesia, *Indonesia Forestry Studies*, Field Document no.I-1, Directorate General of Forest Utilization, Ministry of Forestry/FAO, Jakarta.

Systems Analysis Research Unit (1976), *SARUM 76 global modelling project*, Directorate of Research, Department of the Environment and Transport, London.

Systems Analysis Research Unit (1978), *SARUM Handbook*,

Directorate of Research, Department of the Environment and Transport, London.

Taylor J.E. (1986), Differential migration, networks, information and risk, *Research in Human Capital and Development*, vol.4, pp.147–171.

Thirlwall, A.P. (1976), When is trade more valuable than aid?, *Journal of Development Studies*, no.12, pp.35–41.

Thirlwall, A.P. (1983), Confusion over the relative worth of trade and aid, *World Development*, no.11, pp.71–72.

Thirlwall A.P. (1983), *Growth and development with special reference to developing countries*, London.

Tietenberg T.H. (1988), *Environmental and natural resource economics*, Scott, Foresman and Company, Glenview.

Todaro M.P. (1969), A model of labor migration and urban unemployment in less developed countries, *American Economic Review*, no.59, pp.138–148.

Todaro M.P., Stilkind J. (1981), *City bias and rural neglect: the dilemma of urban development*, The Population Council, New York.

Uhlig H. (1988), Spontaneous and planned settlement in Southeast Asia, in Manshard W., Morgan W.B. (eds), *Agricultural expansion and pioneer settlements in the humid tropics*, the United Nations University, Tokyo, pp.7–43.

UNDP (1990), *Human development report 1990*, Oxford University Press.

UNDP (1994), *Human development report 1994*, Oxford Univesrity Press.

Unesco (1978), *Tropical forest ecosystems*, Paris.

United Nations (1980), *Patterns of urban and rural population growth*, Population studies no.68, Department of Economic and Social Affairs, New York.

United Nations (1986), *Monthly Bulletin of Statistics*, New York.

United Nations (1989), *Prospects of world urbanization 1988*, Population studies no.112, Department of International Economic and Social Affairs, New York.

United Nations (1991a), *National Account Statistics*, New York.

United Nations (1991b), *Statistical yearbook for Latin America and the Caribbean*, 1991 edition, United Nations Publication, New York.

Vergera N.T. (ed.) (1982), *New directions in agroforestry: the potential of tropical legume trees, improving agroforestry in the Asia-Pacific Tropics*, East West Centre and United Nations University, Honolulu.

Vu M.T. (1984), *World population projections 1984, short- and long-term estimates by age and sex with related demographic statistics*, The World Bank, Washington D.C.

Webster C.C., Wilson P.N. (1971), *Agriculture in the tropics*, Tropical Agricultural series, Longman, Harlow, UK.

Wibe S. (1984), *Demand functions for forest products*, IIASA Forest Sector Project Working Paper Wp-84-103, Laxenburg.

Wilkie D.S. (1988), *People of the tropical rainforest*, ed. J. Denslow & C. Padoch, Smithsonian Institute, Washington, pp.111–126.

World Bank (1986), *Indonesia: the challenge of urbanization*, Washington D.C.

World Bank (1990a), *Indonesia, sustainable development of forests, land, and water*, World Bank Country Study, Washington D.C.

World Bank (1990b), *World debt tables 1990–1991*, vol.1, vol.2, supplement, Washington D.C.

World Bank (1992), *World development report 1992*, Oxford University Press, New York.

World Bank (1994), *World development report 1994*, Oxford University Press, New York.

World Resources Institute (1990), *World resources 1990–1991*, Oxford University Press

INDEX